NAVY PLANNING, PROGRAMMING, BUDGETING, and EXECUTION

A Reference Guide for Senior Leaders, Managers, and Action Officers

Irv Blickstein, John M. Yurchak, Bradley Martin,
Jerry M. Sollinger, Daniel Tremblay

Prepared for the United States Navy
Approved for public release; distribution unlimited

For more information on this publication, visit www.rand.org/t/TL224

Library of Congress Cataloging-in-Publication Data is available for this publication.
ISBN: 978-0-8330-9614-2

Published by the RAND Corporation, Santa Monica, Calif.
© Copyright 2016 RAND Corporation
RAND® is a registered trademark.

Cover photo by Ellepistock/iStock.

Support RAND
Make a tax-deductible charitable contribution at
www.rand.org/giving/contribute

www.rand.org

Preface

The Planning, Programming, Budgeting, and Execution (PPBE) process used by the U.S. Department of Defense (DoD) is time-tested and effective, but it is also complex and challenging for new personnel. Furthermore, the process changes in detail relatively frequently, which means that the contextual materials and instruction provided to new action officers become quickly out of date unless they remain at fairly high levels of abstraction. This guidebook responds to a request from the U.S. Navy to produce a reference guide that documents key but enduring aspects of how the Office of the Chief of Naval Operations implements the PPBE process so that action officers, as well as flag officers and senior executives, can successfully navigate and effectively contribute to the process. This guide particularly emphasizes the planning and programming phases because these usually involve the greatest levels of effort and uncertainty, and because they are central to the principal annual deliverable—a coherent, balanced Program Objective Memorandum (POM) in alignment with leadership's guidance. As this report was being prepared for publication, the RAND research team became aware that the Navy is planning some changes to its POM process. This guide is accurate as of December 2015.

The guidebook is aimed at action officers, branch heads, newly assigned flag officers and executives in the Office of the Chief of Naval Operations, and outside stakeholder organizations with an interest in the PPBE process and how the Navy executes it.

This research was sponsored by the Office of the Chief of Naval Operations, Programming Division (N80) and was conducted within the Acquisition and Technology Policy Center of the RAND National Defense Research Institute, a federally funded research and development center sponsored by the Office of the Secretary of Defense, the Joint Staff, the Unified Combatant Commands, the Navy, the United States Marine Corps, the defense agencies, and the Defense Intelligence Community.

For more information on the RAND Acquisition and Technology Policy Center, see www.rand.org/nsrd/ndri/centers/atp or contact the director (contact information is provided on the web page).

Contents

Figures

Summary

Policies and procedures that cover how the Office of the Chief of Naval Operations (OPNAV) staff participates in the U.S. Department of Defense (DoD)'s Planning, Programming, Budgeting, and Execution (PPBE) process have been documented to varying degrees over the years. This documentation typically has taken several forms, each with different objectives. First, there are various policy and procedural documents associated with preparing, submitting, and defending the annual Program Objective Memorandum (POM) submission and its supporting documentation. Second, there are policy and procedural documents associated with properly interacting with the various PPBE-related information systems that centralize and mechanize key aspects of the PPBE process and its key deliverables and supporting products, including the POM database. Third, there are training materials designed to familiarize new OPNAV staff with the background, purpose, and context of PPBE. The first two forms of documentation are essentially technical in nature—for example, procedures, definitions, stakeholders, rules, schedules, deliverables, specifications, and formats. The third form of documentation is a necessarily brief introduction to all phases of the PPBE process, the major stakeholders and their roles, the typical PPBE process flow and schedule, and the major products developed and delivered throughout a typical year.[1]

OPNAV's Programming Division (N80) identified the need for an authoritative reference guide that would fill the gap between the technical procedural documentation and the courseware. Division leaders saw the need for a guide that documents key PPBE processes, decision points, products, procedures, and relationships that AOs should understand to be effective and successful in the PPBE process. The leaders also wanted to provide a reference for more-senior leaders and managers in key stakeholder organizations that either contribute to or depend on a coherent, balanced POM. N80 asked the RAND Corporation's National Defense Research Institute to develop such a reference guide. This document is RAND's response to that request.

---

[1] The action officer (AO) course is a two-day introduction to PPBE. Flag officers and Senior Executive Service members get a one-day executive overview.

In approaching this task, the RAND team first identified the historical and refer-ence documents that would provide authoritative sources for such a guide. The team then developed a story line for the guide and interviewed subject-matter experts with extensive experience in PPBE both inside and outside of OPNAV. The resulting guide-book is aimed at AOs, branch heads, newly assigned flag officers and executives in OPNAV, and outside stakeholder organizations with an interest in the OPNAV PPBE process and how the Navy executes it. The guide tries to represent the perspective of requirements and program advocates, as well as requirements and resource integrators. It is intended as an adjunct to the technical procedures and course material offered by OPNAV, focusing on how PPBE actually works within the Navy. It tries to highlight those aspects of the PPBE process, products, stakeholders, relationships, interactions, and best practices needed for readers and their organizations to be successful individu-ally and as a staff working toward the common purpose of delivering a coherent, bal-anced POM in alignment with leadership guidance.

Key Takeaways

The PPBE process overlaps with two other key DoD processes: the Defense Acquisi-tion System and the Joint Capabilities Integration and Development System. These three complex processes interact with each other and cannot be understood in isola-tion. A requirements decision triggers an acquisition decision, which creates a demand for resources. Curtailing resources affects acquisition and may affect the ability to meet a requirement. Moreover, numerous organizations have equities in PPBE and either participate directly or depend on its outcomes. Failing to inform and coordinate with stakeholders can create ill will among organizations and generate significant addi-tional staff work, which often is performed under time pressures. Also, the output of the Navy's PPBE process is an input into the Department of the Navy (DoN)'s PPBE process,[2] whose output is an input into DoD's PPBE process. Understanding the per-spectives of all of these organizations is a prerequisite to success.

Key phases and decision points in the PPBE process include strategic planning, the issuance of guidance, requirements assessment, the building and integration of the POM, POM review (to determine whether it complies with guidance), end game (or final decision process), and budget review.

Several key stakeholders play a role in OPNAV's PPBE process. These include the CNO, who owns the Navy POM; the Deputy Chief of Naval Operations for Integra-tion of Capabilities and Resources (N8); its Programming Division (N80); its Assess-ment Division (N81); resource sponsors, who define, advocate for, and defend specified

---

2   The Navy PPBE process and DoN PPBE process are not the same. The Chief of Naval Operations (CNO) and Commandant of the Marine Corps each submit a POM to the Secretary of the Navy, who then must submit a balanced department POM to the Office of the Secretary of Defense.

subsets of the Navy's capability requirements; and requirements sponsors, who highlight issues of collective importance but do not necessarily have programming authority over the resources involved. Understanding where one fits in the system of roles and relationships between these stakeholders is a prerequisite to effective participation in the PPBE process.

A large number of other stakeholders play directly or indirectly in the PPBE process. They range from the Fleet Commanders and Type Commanders to the materiel establishment, including the Systems Commands, Program Executive Offices, and warfare centers. Also critical are members of the Navy Secretariat, including the Secretary of the Navy (who owns the department POM) and his or her deputy; the Assistant Secretary for Research, Development, and Acquisition; and key members of the U.S. Marine Corps staff. Key players outside the Navy have major and direct roles as well. These include the Chairman of the Joint Chiefs of Staff; Congress; and members of the DoD staff, such as the Office of the Under Secretary of Defense (Comptroller), the Office of the Under Secretary of Defense (Acquisition, Technology, and Logistics), the Office of Cost Assessment and Program Evaluation, and the Secretary and Deputy Secretary of Defense.

At the AO level, being successful rests on the following common-sense principles:

1. Understand that the Navy's POM belongs to the CNO. Thus, it becomes critical to understand the CNO's vision and priorities for the POM and how he or she plans to interact with his or her senior leaders and staff to deliver the POM to the Secretary of the Navy. Gaining that understanding requires sustained effort over time—both because the vision may not be spelled out in detail or in one place and because how the CNO chooses to do business and make decisions may change over time. Together, guidance, speeches, congressional testimony, and the CNO's past decisions can indicate what that vision and approach might be.
2. Understand that your program or portfolio of programs is but one of many. It is critical to know both the overall Navy Program and where your portfolio fits within it.
3. Have a plan for how to justify and manage the movement of your portfolio through the various phases of the process, as well as how to adapt this plan to changes as they surface.

To summarize, the OPNAV PPBE process has many claimants, but there are some common themes to success. The authors of this reference guide, who have held various positions in OPNAV and together accumulated more than 50 years of experience and many POMs, observe that a successful POM

*   is on time
*   is integrated—vertically and horizontally

- unambiguously expresses the CNO's priorities
- aligns with the CNO's guidance on balancing requirements, resources, and risk
- is coherent with the CNO's POM story line (or theme).

These and other best practices outlined in the guidebook can facilitate and improve Navy staff's ability to work together toward delivering a coherent, balanced POM in alignment with leadership guidance.

Acknowledgments

The authors would like to express appreciation for the excellent advice and counsel received from the guidebook's three reviewers, Stephanie Young, Christopher Mouton, and Charles Nemfakos. All are experienced professionals with years of direct exposure to the U.S. Department of Defense's Planning, Programming, Budgeting, and Execution processes, particularly the Navy's process. We have taken their comments to heart, and the document is much improved by their efforts.

Abbreviations

ACAT	acquisition category
AO	action officer
ASN	Assistant Secretary of the Navy
AT&L	Acquisition, Technology, and Logistics
BAM	Baseline Assessment Memorandum
BSO	budget submitting office
C4ISR	command, control, communications, computer, intelligence, surveillance, and reconnaissance
CAPE	Office of Cost Assessment and Program Evaluation
CDD	Capability Development Document
CNO	Chief of Naval Operations
CPD	Capability Production Document
DCNO	Deputy Chief of Naval Operations
DoD	U.S. Department of Defense
DoN	Department of the Navy
FEA	Front End Assessment
FLTCOM	Fleet Commander
FM&C	Financial Management and Comptroller
FMB	Financial Management and Budget
FY	fiscal year

FYDP	Future Years Defense Program
HQMC	Headquarters Marine Corps
ICD	Initial Capabilities Document
J8	Force Structure, Resources, and Assessment Directorate
JCIDS	Joint Capabilities Integration and Development System
JROC	Joint Requirements Oversight Council
KPP	key performance parameter
N1	Deputy Chief of Naval Operations for Manpower, Personnel, Training, and Education
N2/N6	Deputy Chief of Naval Operations for Integration Dominance
N2/N6F	N2/N6 Warfare Integration Directorate
N3/N5	Deputy Chief of Naval Operations for Operations, Plans, and Strategy
N4	Deputy Chief of Naval Operations for Fleet Readiness and Logistics
N8	Deputy Chief of Naval Operations for Integration of Capabilities and Resources
N9	Deputy Chief of Naval Operations for Warfare Systems
N9I	N9 Warfare Integration Directorate
N80	Programming Division
N81	Assessment Division
N82	Fiscal Management Division
N84	Innovation, Technology Requirements, and Test and Evaluation Division
N89	Special Programs Division
N95	Expeditionary Warfare Division
N96	Surface Warfare Division
N97	Undersea Warfare Division
N98	Air Warfare Division

N99	Unmanned Warfare Systems Division
N801	Program Planning and Development branch
N802	Integration and Alignment branch
N803	Navy, Joint, and Urgent Requirements branch
NDAA	National Defense Authorization Act
NSP	Navy Strategic Plan
OPNAV	Office of the Chief of Naval Operations
OSD	Office of the Secretary of Defense
OUSD	Office of the Under Secretary of Defense
P&R	Programs and Resources Division
PEO	Program Executive Office
PM	program manager
POM	Program Objective Memorandum
PPBE	Planning, Programming, Budgeting, and Execution
R3B	Resources and Requirements Review Board
RDA	Research, Development, and Acquisition
RO	requirements officer
SECNAV	Secretary of the Navy
SPP	Sponsor Program Proposal
SYSCOM	Systems Command
TYCOM	Type Commander
USMC	U.S. Marine Corps
VCJCS	Vice Chairman of the Joint Chiefs of Staff
WSCA	Warfighting and Support Capability Assessment

CHAPTER ONE

Introduction

Policies and procedures that cover how the Office of the Chief of Naval Operations (OPNAV) staff participates in the U.S. Department of Defense (DoD)'s Planning, Programming, Budgeting, and Execution (PPBE) process have been documented to varying degrees over the years. This documentation typically has taken several forms, each with different objectives. First, there are various policy and procedural documents associated with preparing, submitting, and defending the annual Program Objective Memorandum (POM) submission and its supporting documentation. Most of these are updated annually because they usually define the schedule, specifications, and guidance for developing the annual deliverables and decision-support products involved in the PPBE process. Second, there are policy and procedural documents associated with properly interacting with the various PPBE-related information systems that centralize and mechanize key aspects of the PPBE process and key deliverables and supporting products, including the POM database. These are updated less frequently—normally, when features, specifications, or policies related to these systems or the products or data they operate on change. Third, there are training materials designed to familiarize new OPNAV staff with the background, purpose, and context of PPBE. The first two forms of documentation are essentially technical in nature—for example, procedures, definitions, stakeholders, rules, schedules, deliverables, specifications, and formats. The third form of documentation is a necessarily brief introduction of all phases of the PPBE process, the major stakeholders and their roles, the typical PPBE process flow and schedule, and the major products developed and delivered throughout a typical year.[1]

The challenge is that the PPBE process changes frequently from year to year, driven by changes in leadership—for example, Service staffs, Office of the Secretary of Defense (OSD) staffs, and elected officials—as well as changes in fiscal and political environments. The technical documentation generally keeps up with and adapts to these changes because the authors of such documents are mainly the data or process owners. But the training courseware cannot keep up. This is not because the PPBE process changes too frequently but because the courseware must be an overview aimed at

---

[1] The action officer (AO) course is a two-day introduction to PPBE. Flag officers and Senior Executive Service members get a one-day executive overview.

PPBE novices. Trying to describe subtle but important process or product changes in a one- or two-day course to members still trying to learn the abbreviations and acronyms is not an effective way to spend anyone's time. It takes a complete annual POM cycle of on-the-job training before a typical AO understands the purpose of the big PPBE process and product pieces. It typically takes a second POM cycle for an AO to begin to develop the "journeyman" understanding and skills needed to be effective as a requirements officer (RO) or programmer. During these cycles, the technical documentation is always there as a ready reference to reinforce proper procedures and the mechanics of how the products combine to develop the POM submission. What is missing is a guide for how to be successful in the process. In the absence of such a guide, *success* is often defined by where and for whom a staff member works.

This guidebook is aimed at AOs, branch heads, newly assigned flag officers and executives in OPNAV, and outside stakeholder organizations with an interest in the OPNAV PPBE process and how the Navy executes it. The approach followed in writing the guide was to first identify the historic reference documents that would provide authoritative sources for such a guide. The RAND Corporation team then developed a story line and interviewed subject-matter experts (both current and former OPNAV Senior Executive Service members) with extensive experience in PPBE both inside and outside of OPNAV. This guide tries to represent the perspective of requirements and program advocates, as well as requirements and resource integrators. It is intended as an adjunct to the technical procedures and course material offered by OPNAV, focusing on how PPBE actually works within the Navy. The guide tries to highlight those aspects of the PPBE process, products, stakeholders, relationships, interactions, and best practices needed for readers and their organizations to be successful individually and as a staff working toward the common purpose of delivering a coherent, balanced POM in alignment with leadership guidance.

What OPNAV Does

According to the OPNAV *Organization and Operations Manual* (OPNAV, 2011), the OPNAV staff exists to perform the following 11 key functions:

1. Field a Naval force capable of carrying out tasking from higher authority.
2. Investigate and report on Navy readiness.
3. Establish Navy strategy and policy, and issue guidance.
4. Align actions of Navy organizations.
5. Plan and program in support of the POM.
6. Plan and coordinate Navy employment.
7. Translate maritime strategy into strategic guidance and priorities.
8. Integrate requirements.

9. Determine fiscal distributions and allocations.
10. Conduct operational test and evaluation.
11. Conduct and manage all aspects of intelligence assessment throughout the Department of the Navy (DoN).

This guide essentially is about the fifth function, but several others involve much of the work and decision-support products needed to facilitate a coherent, balanced POM.

How This Guide Is Organized

Chapter Two provides an overview of the PPBE process. It includes a brief discussion of the key processes in the acquisition cycle and how they interrelate, as well as a historical perspective from the Navy's point of view. It also briefly discusses the processes in DoD and the U.S. Marine Corps (USMC).

Chapter Three describes how OPNAV carries out its PPBE process. It describes the key functional roles, the Navy's PPBE processes and timelines, and the key phases, decision points, and deliverables.

Chapter Four describes the key internal and external stakeholders, the roles they play in the PPBE process, and important relationships among them.

Chapter Five catalogues best practices and techniques for contributing successfully to the development of the POM and the ensuing budget.

Chapter Six provides advice for how to plan for effective participation and how to maximize the potential for successful outcomes in the PPBE process.

Chapter Seven offers concluding thoughts.

How to Use This Guide

Whether assigned as an RO or programmer in a resource sponsor, as a programmer in OPNAV's Programming Division (N80) (whose workspace is known as the "bull-pen"), or as an analyst in OPNAV's Assessment Division (N81), each new AO on the Navy staff should read this handbook cover to cover after attending the two-day PPBE course to gain more-nuanced understanding of how the POM and budget processes work in OPNAV. AOs should retain the guide as a desk reference as they participate in building and defending the POM.

Leaders new to the Pentagon should review the guide after taking the one-day PPBE course, and those outside the Pentagon can use it as a quick reference to understand OPNAV's processes by referring to specific topics of interest from the table of contents.

CHAPTER TWO

PPBE Process Overview

The PPBE process is one of three overlapping resource decision processes in DoD. Many other influences, from events overseas to actions taken by Congress or other executive agencies, can affect these processes. In this chapter, we describe the three resource decision processes, summarize them within the PPBE process, and discuss additional important external influences.

The Three Key DoD Decision-Support Systems

DoD uses three interrelated systems to ensure correct documentation of requirements, oversight of acquisition, and proper resource levels for acquisition and all other department activities. The three systems are the Joint Capabilities Integration and Development System (JCIDS), the Defense Acquisition System, and the PPBE process. The Defense Acquisition University describes this process with a Venn diagram, shown in Figure 2.1.

The JCIDS process supports the Joint Requirements Oversight Council (JROC) and the Chairman of the Joint Chiefs of Staff by identifying, assessing, validating, and prioritizing joint military capability requirements.[1] The JCIDS provides a transparent process that allows the JROC to balance the demands of the uniformed Services and make informed decisions when validating and prioritizing capability requirements.

The Defense Acquisition System takes validated capability requirements and turns them into materiel capability solutions. JCIDS documents link validated capability requirements and the acquisition of materiel capability solutions through five major Defense Acquisition System phases; these phases and the acquisition process are depicted in Figure 2.2. It is through this process that OPNAV staff comes in direct contact with Program Executive Offices (PEOs), program managers (PMs), Systems Commands (SYSCOMs), and their warfare centers.

PPBE is the primary process for enabling the funding of the various JCIDS and Defense Acquisition System activities, including operations and support that develop,

---

[1] The descriptions of the JCIDS, Defense Acquisition System, and PPBE process are drawn from Defense Acquisition Portal, undated-b.

Figure 2.1
DoD Decision-Support Systems

SOURCE: Defense Acquisition Portal, undated-a.
RAND *TL224-2.1*

Figure 2.2
Acquisition Process

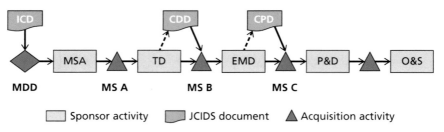

SOURCE: Defense Acquisition Portal, undated-b.
NOTE: CDD = Capability Development Document; CPD = Capability Production
Document; EMD = Engineering and Manufacturing Development; ICD = Initial
Capabilities Document; MDD = Materiel Development Decision; MS = Milestone;
MSA = Materiel Solution Analysis; O&S = Operations and Support;
P&D = Production and Deployment; TD = Technology Development.
RAND *TL224-2.2*

field, and sustain effective capability solutions for the warfighters. Understanding the relationships among the three DoD decision-support systems is important to success in any one of them. PPBE alone cannot ensure effective requirements development or defense acquisition effectiveness. It can undermine both, however, if resources are not made available to fund requirements or allow economical acquisition. Similarly, poorly understood requirements or acquisition programs can cause unexpected demands on the resource process, forcing difficult choices between current and future capability, support of operating forces, and acquisition. Worth noting is that the drivers of these processes do not necessarily coincide. The PPBE process is calendar-driven, whereas the requirements and acquisition systems are event-driven—for example, by a milestone decision or the emergence of a capability gap.

Historical Perspective

Although the DoN has been budgeting, developing, and defending needs and procuring equipment throughout its existence, most of the processes and systems described in this guide are comparatively recent. PPBE traces its origin back to the tenure of Robert McNamara, Secretary of Defense under Presidents John F. Kennedy and Lyndon B. Johnson. Having a background in analysis and having managed a large corporation, McNamara's goal was to rationalize what he viewed as the chaotic Service processes that had been used before his tenure. Remarkably, although the PPBE system has varied in some steps and players since its inception, it has remained the dominant DoD resource-planning process without major changes.

The requirements process for large programs is currently governed by the JCIDS, which originated in the early 2000s under the direction of James Cartwright, then–Lieutenant General of the Force Structure, Resources, and Assessment Directorate (J8) and, later, Vice Chairman of the Joint Chiefs of Staff (VCJCS). The JCIDS is an internal DoD process, but it is designed to support the statutory JROC, formed as a result of the 1986 Goldwater-Nichols Act. The main purpose of the JROC—and its supporting system, the JCIDS—is to inject more rigor into the capability development process, particularly to ensure that capabilities are "born joint." Each Service has requirements systems that support the JROC and, in many important ways, emulate its structure and decisionmaking.

The Defense Acquisition System has its origins in the 1970s Packard Commission and has undergone numerous, mostly superficial, revisions. The Weapon Systems Acquisition Reform Act of 2009 did impose statutory changes with direct influences on the system and its relationships with the other elements of resource development.

Interrelationships

PPBE both influences and is influenced by the other decision-support systems. Not having the same drivers as PPBE, the JCIDS and Defense Acquisition System processes have the potential to produce requirements that those developing the POM might not expect. Unexpected requirements can lead to churn, underresourcing of programs, and disruptions to other processes. Ultimately, the goal of PPBE is to deliver balanced, executable programs and budgets.

While many of the players in the PPBE and JCIDS processes are the same—or at least work in the same offices—there is sometimes a tendency to drop curtains between requirements, programming, and acquisition. Acquisition has been the responsibility of the Navy Secretariat,[2] through the SYSCOMs, PEOs, program offices, and supporting warfare centers. Clearly, such agencies and entities need insight into program funding expectations and expected shifts in requirements. When this does not take place, essential decisionmaking information might be missed. Likewise, those managing the Navy Program need visibility on acquisition issues as early as possible to avoid last-minute or late funding misalignment.[3]

The programming and budgeting processes have been affected by congressional efforts to reduce the federal deficit, particularly through the Budget Control Act of 2011. While the PPBE process is capable of continuing with reduced resource levels, the uncertainty in timing and funding levels has disrupted the process, sometimes requiring the preparation of multiple POMs or budgets, or both, in the same year.[4] Moreover, limits imposed have led to funding levels unrelated to any figure that a systematic resource-planning process would generate. This is not to say that the Navy and other Services should not generate resource plans, but political actions intentionally designed to tie the hands of government agencies will be in play.

OPNAV PPBE Process

OPNAV's PPBE is a resource-allocation process intended to rationally apply leadership priorities and ensure appropriate support levels for all of the Navy's activities. It affects the acquisition process, but it also affects and is affected by current and future operations, personnel policy, and overall readiness goals. When completed correctly, it ensures that acquisitions and requisite support programs are properly funded, that current priorities are sustained, and that the Navy has made the best—or, increasingly, the least bad—choices given available resources. Moreover, a well-formulated POM, at minimum, provides a defensible rationale for the Service's overall requirements.

Figure 2.3 is drawn from the OPNAV PPBE course for AOs and shows the major phases of the DoD PPBE process along the timeline for the POM whose programs will

---

[2]   As this guide was being written in 2015, members of Congress sought to shift some acquisition authority from OSD and Service Secretaries to the Service Chiefs, which resulted in specific language in the fiscal year (FY) 2016 National Defense Authorization Act (NDAA). Services have responded to that language, but the final disposition of these authority shifts remains to be seen.

[3]   The *Navy Program* essentially means the sum of all of the individual acquisition programs within the Navy, plus manpower funding and readiness funding.

[4]   Multiple POMs usually arise as a result of leadership intentions to develop one POM based on more-favorable or optimistic topline assumptions, as well as one or more other POMs to "hedge" positions based on more-pessimistic (or realistic) assumptions about the resources that may or will more likely be approved in the NDAA.

Figure 2.3
U.S. Department of Defense PPBE Timeline

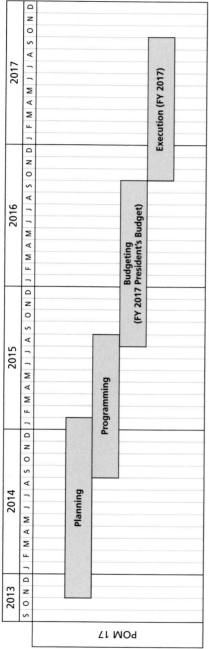

SOURCE: DoN, 2015a, Block I.

RAND TL224-2.3

be executed in FY 2017 ("POM17"). Each phase takes about a year, but phases can overlap. So, for programs that will be executed in FY 2017, the planning must begin in FY 2013. In this section, we describe the basic elements and timelines of each of the four phases: planning, programming, budgeting, and execution. Later chapters discuss the phases in more detail and make suggestions for successful participation and contribution in each.

Planning Phase

At some level, all OPNAV planning activities impinge on PPBE. Some descriptions of the PPBE process describe these planning activities as integral, but they are, in fact, general influences rather than direct inputs. As a practical matter, the inputs that come closest to planning in PPBE are those developed by (1) resource sponsors to advance or align the capabilities and capacity of their program portfolios to existing or emerging requirements and (2) N81, which assesses the capability and program plans of the sponsors in the broader context of Fleet or Enterprise activities. In this case, *planning* means relating a set of needs and shortfalls to a set of resource priorities, with relationships developed to the point that more-detailed programming of resources can occur. In fact, the Defense Planning Guidance and the Navy Strategic Plan (NSP) all influence the character of the POM.

The program of record will generate a set of capabilities intended to execute a set of missions and plans in various operational environments. Some elements of the program may be insufficient to meet an emerging threat; on the other hand, they may exceed minimal acceptable capability. Moreover, acquisition elements in the program may have fallen behind schedule, and the resulting capability gaps may require mitigation. Not every shortfall is a matter of immediate interest. Planning allows the assessment of risk associated with particular gaps. OPNAV N81 provides Navy-wide independent assessments and, through its analytic efforts, advises leadership on priorities and potential trade-offs, consequences, and risks. Other players provide input into the Navy-wide assessments while performing more-narrow assessments for their own warfare areas.

Programming Phase

Programming identifies available resources, prioritizes needs across the Future Years Defense Program (FYDP), and then outlines these for inclusion in the POM, which serves as both a resource guide and a statement of priority to internal and external audiences over a five-year period. The main deliverable in this phase is the POM resource database, which is accessed and manipulated through dedicated enterprise software called the Program Budget Information System.

The planning phase will have identified capability needs, gaps, and possibly excesses and will have assigned a degree of risk associated with the projected consequences of each shortfall. It is the particular business of the programming phase to balance the correction of shortfalls given available resources. For major programs, it provides a medium-term resource plan, and for the whole Navy, it provides a medium-term plan for materiel and personnel sustainment.

OPNAV's N80 is principally responsible for preparing the POM and does so largely under the direct guidance of the Chief of Naval Operations (CNO), via the Deputy Chief of Naval Operations (DCNO) for Integration of Capabilities and Resources (N8). However, N80 uses Sponsor Program Proposals (SPPs) as a basis, as well as the planning and assessment documents provided by N81. Because the POM ultimately leaving the DoN is a product of both the Navy and USMC, N80 regularly interacts with the Navy Secretariat and ideally interacts regularly with its USMC counterpart during the programming phase.

Because the POM is a five-year statement of intent, portions past the first year or two are less certain to reach the execution phase than they are to reach budget submission. While it is more specific than planning recommendations, the POM should be viewed as a political document as much as a fiscal one, conveying the CNO's intentions, priorities, and vision for his or her tenure. CNOs cannot necessarily influence decisions that might need to be made ten years in the future. They can, however, directly influence budget decisions and strongly influence decisions made within the timing of the FYDP. This guide will frequently return to the necessity of aligning POM priorities with those of the CNO. Of the fiscal documents that OPNAV produces, none has as close an association with the CNO as the Navy POM.

For resource sponsors, it is important to recognize the POM's significance to particular programs. Although having an entry in the POM does not guarantee out-year support, failing to appear or having underfunded support in the POM telegraphs a priority that does not bode well for the program's treatment at the hands of OSD or Congress. If a sponsor has not convinced N80 of the importance of support, the sponsor may need to rethink how the program is planned or how its requirements are communicated.

Both programming and the next phase, budgeting, are financial processes involving the allocation of money. They do, however, have separate time horizons, deliver different products, and interact with different OSD offices. Historically, until the mid-2000s, program and budget reviews within DoD occurred sequentially, first with the POM submission, then a program review held at the OSD Office of Program Analysis and Evaluation. Thereafter, military department budgets were submitted and then reviewed by the Office of the Under Secretary of Defense (OUSD) (Comptroller).

But since 2004, program and budget reviews have been held simultaneously. While the differing roles of the Office of Program Analysis and Evaluation (and now its successor, the Office of Cost Assessment and Program Evaluation [CAPE]) and the OUSD (Comptroller) are recognized, the processes are combined at the "end game," which has made it difficult for senior departmental leaders to ascertain the difference.[5] As of 2015, there are some indications that program and budget reviews could be separated and treated again as sequential processes. However, whether these indications will translate into a process change is not yet clear.

From the perspective of those operating in the PPBE process, it is important to understand the things that must be funded in an upcoming execution year and the things that need to be recognized as priorities across the FYDP. It is also important to ensure that items appearing in programming and budgeting documents are consistent so that no mixed messages are transmitted to internal or external audiences.

Budgeting Phase

If programming is an organizational plan and statement of priority, *budgeting* is the enterprise acting within legal (i.e., fiduciary) constraints reflecting the will of Congress. POMs can be somewhat imprecise in the out-years of the FYDP, but budgets reflect an actual amount of funding available for operations, procurement, personnel support, and all the other activities. Budgets are thus more precise, especially in the budget year.

Some OPNAV codes, such as the Fiscal Management Division (N82) and Innovation, Technology Requirements, and Test and Evaluation Division (N84), are dual-hatted as agencies in both OPNAV and the Navy Secretariat, which is the ultimate authority for DoN's budget. In this capacity, OPNAV N82 is dual-hatted as the Assistant Secretary of the Navy (ASN) for Financial Management and Budget (FMB) office. While the interests of OPNAV and the broader DoN generally converge, they are not identical, and it is important for stakeholders to understand this. In addition, headquarters-level budgeteers regularly interact with budget submitting offices (BSOs) in those Navy field activities whose priorities are shorter term and may or may not conform to the CNO's longer-term vision or with resource sponsor priorities.

Execution Phase

While budgeting is generally associated with the preparation and presentation of a budget, *execution* refers to the actual allocation and expenditure of funds. While this process might not immediately concern an OPNAV staff member focused on planning and programming, it is of great interest to entities with spending and daily operational responsibilities, such as Fleets or program offices. In general, the nearer the process is to actual expenditure of money, the more restrictive the controls and the greater the need for precision. Actions taken in other processes that may make current-year spending difficult to execute—for example, abrupt changes in the near-term program profile—may induce uncertainty and forced inefficiency in spending. Actors in other parts of the process should be mindful of these potential effects. Ultimately, resource sponsors must at least be aware of execution activities because they involve the Enterprise

---

[5]   *End game* is the term commonly used for the final balancing of accounts. See Chapter Three.

or Fleet capabilities in operation and because issues related to these operations have implications for existing or emerging requirements that signal demand for resources.

Requirements Development and Approval: The JCIDS and Related Processes

The term *requirement* means different things to different audiences and is thus subject to abuse and misunderstanding. A requirement can be the totality of resources needed to accomplish some mission or function. It can be the unconstrained, as opposed to the funded, need. It can also be things critically needed as opposed to things simply desired. Given how important the idea of a requirement is to what OPNAV does in general and to the PPBE process in particular, it is surprising how often the term is used in the absence of specific context. In the Navy and U.S. Air Force, a requirement is nominally defined as the capabilities or conditions possessed by a system or a component needed to solve a problem or achieve an objective. This definition works for PPBE as well, but it needs a qualifying term for its meaning to PPBE to be understood. The key point here is that, unless used in conjunction with an adjective or adjunct noun, the term *requirement* is ambiguous. Here are examples of the specific types of requirements that stakeholders need to understand (and differentiate) to be successful in the PPBE process:

- *Capability requirement*: In the JCIDS and OPNAV requirements assessment processes, this term refers to either (1) the capabilities to achieve military objectives identified in a mission, campaign, or capabilities-based assessment (as documented in an ICD) or (2) the operational performance attributes at a system level necessary for the acquisition community to design a proposed system and establish a program baseline (as described in a CDD or CPD). This type of requirement is also often referred to as a "warfighting" or "operational" requirement.
- *Capacity requirement*: This usually comes up in the context of sufficiency analysis (such as OPNAV N81 or Joint Staff campaign modeling, ordnance modeling, or mission-level modeling or analysis, or during N81 force-structure assessments). It describes the assessed need for numbers of types and configurations of platforms and/or systems within a given operational or mission context to achieve one or more military objectives (such as neutralizing a missile raid or eliminating a target set within a specified time).
- *Performance requirement*: Generally, such requirements are the operational performance attributes at a system level necessary for the acquisition community to design a proposed system and establish a program baseline (as described in a CDD or CPD).

- *Program (or programmatic) requirement*: Usually this is (1) a specific program objective to satisfy a validated capability or capacity requirement within a certain time frame, (2) an acquisition objective to meet a certain execution profile, or (3) the resources needed to achieve (1) or (2) within a certain time frame.
- *Resource requirement*: The resources needed to satisfy a specified capability, capacity, performance, or program requirement.
- *Validated requirement*: In reference to requirements, the terms "Big R" and "little r" refer to whether a requirement has been validated by a recognized decision authority, such as the Resources and Requirements Review Board (R3B) or the JROC. Generally, "Big R" requirements are JROC- or R3B-validated requirements. Validation is essentially an entry fee to be qualified to compete for resources in the POM (i.e., sponsors should not be resourcing nonvalidated requirements). What constitutes validation varies by context and should be understood before committing resources. "Little r" requirements, on the other hand, are any non-validated expression of need.
- *Whole requirement*: This term generally comes up in the context of accounts or broad sets of requirements and is almost always across numerous programs and sponsors. For example, the Navy's manpower account, various readiness accounts, infrastructure accounts, and maintenance accounts all resource multiple program requirements. When resource or platform sponsors assert program-specific requirements, the N8 will frequently direct a requirements sponsor to perform a consolidated assessment of these (capability, capacity, performance, or program) requirements to determine the resources needed to meet all (or some predefined percentage) of them. This is a major objective of a Baseline Assessment Memorandum (BAM).

The requirements process has a specific relationship to capability development and generally involves new capabilities. The Navy has had an R3B for many years. It is normally chaired by the N8 and, depending on the decision, the Assistant Secretary for Research, Development, and Acquisition (RDA). The board exists, among other reasons, to validate capability requirements submitted by Fleets or resource sponsors. The R3B delegates some of this decisionmaking to the Navy Capabilities Board for capabilities of interest—but not capabilities of such interest that a three-star review is deemed necessary.

A joint requirements process can, at times, significantly affect the Navy, including the PPBE process. The JROC is a four-star body intended to ensure that the requirements for all major programs are reviewed and that the capabilities satisfy not just Service needs but those of combatant commanders. It is intended to promote interoperability, reduce duplication, and ensure attention to the most pressing capability gaps. Every program involving capability development receives a review by a "gatekeeper" in the J8 of the Joint Staff, who assigns it a joint potential designator. Many programs are deemed "Service interest," and requirements approval responsibility remains with the initiating Service or agency. Others receive joint potential designators that direct them to the JROC. In practice, most of the items deemed to require JROC review are actually reviewed and approved at lower levels in the JROC structure. However, all will go through the JCIDS process.

The JCIDS influences the capability development process through the following documents:

- The ICD formally documents gaps identified in capabilities-based assessments and recommends possible material solutions. This forms the basis for an Analysis of Alternatives—a process that recommends a specific material solution from among several potential alternatives.
- A CDD describes a material solution and specifies key performance parameters (KPPs), as well as key cost and schedule parameters. These are the critical capabilities that the program must achieve and demonstrate during operational testing. Changing KPPs requires permission from the approval authority.
- A CPD revalidates KPPs and lays out a solution production phase and delivery schedule.
- A Doctrine, Organization, Training, Leadership, Personnel, and Facilities Change Review documents nonmaterial solutions to capability gaps that should be considered as alternatives to or in addition to a material solution.

All of these documents require a sometimes lengthy approval process. The JCIDS has come under criticism for its cumbersome, document-focused approach and has been revised several times to truncate some processes or satisfy urgent needs. However, Navy internal processes are of equal concern.

From the perspective of participants or stakeholders in the PPBE process, the requirements generation system creates several effects to consider. First, requirements cost money. Some have suggested that actors in the JROC are inclined to identify requirements but are disinclined to trade any away. The result is that, essentially, any program comes to the program and budgetary process with requirements in excess of resources. Repeated efforts to constrain the requirements process have not yet brought requirements more in line with fiscal reality, and the result is either smaller quantities than requested or hasty choices made by programmers with tight deadlines and relatively little information to inform decisions.

There has always been a debate between requirements advocates and resource integrators about whether fiscal constraints should be part of the requirements generation process. This guide does not take a position in that debate, but it can offer lessons learned over the past few decades. First, there *is* value in knowing what it will take to accomplish a mission or objective independent of resource constraints. However, the requirements process is usually constrained by other pragmatic factors that already moderate the result to some degree. Most of these factors relate to the simple fact that future capability requirements usually *must* evolve from existing capabilities or take present or projected capabilities into account. For example, a completely unconstrained requirement that suggests a platform or weapon that is physically incapable of operating with existing or programmed platforms is likely to be changed to ensure interoperability.

This self-constraint is already part of the requirements development process. The more challenging aspect of unconstrained requirements is when two or more must compete for insufficient resources. As difficult a decision as this may be, when requirements advocates do not make the effort to frame competing requirements within some sort of prioritization scheme, they essentially hand the trade decision *and* their voice in the requirements process to the resource integrator. In other words, in the absence of a coherent capability or program-planning rationale for preferring one requirement over another when resources are insufficient, the decision may end up being fiscally driven.

When a requirement is established, there is an expectation from the Secretary of Defense that the sponsoring Service will attempt to meet it. Thus, an important aspect of the requirements process is planning for whether and how it could be resourced if validated and approved. Each new requirement represents a new bill against an existing pool of resources. If a Service does not adequately fund a priority program, the Secretary of Defense (or, more likely, the Deputy Secretary of Defense or the Director of CAPE) may direct that it be funded, which will affect other parts of the overall Service program. Services are well advised to review requirements carefully before they become established and be prepared to defend decisions to defer or simply not fund a requirement.

The Joint Staff also plays a more direct role in the PPBE process as a result of its annual attempts to gauge the degree of congruence between Service POM submissions and the capability priorities established by the JROC (there is some congruence between the Joint Chiefs of Staff assessment and a similar one performed by N81). While the assessment itself cannot compel a change, it may provide the basis for a change directed by OSD.

Defense Acquisition System

The Defense Acquisition System is a complicated process that translates requirements into capabilities. It involves selecting suppliers, contracting for elements and subelements of systems, engineering components and subcomponents, and performing test and evaluation. The actors in this system regularly interact with those in the requirements and PPBE communities, but in the Navy, the majority of personnel report to ASN (RDA) rather than to the CNO.

This system has numerous statutory and regulatory requirements. Decision points are event-driven, and typical events are achievement of or delay in a milestone. Acquisition works best when funding is known and consistent. This need for known funding can and frequently does conflict with the requirement for balanced programs and budgets. In fact, the need to use program funding to achieve balance—while understandable—can wreak havoc on programs that are performing well. Conversely, protecting funding for programs that are not performing well may be suboptimal from multiple standpoints. Some programs experiencing technical delays may, in fact, be appropriate for deferral. Monitoring the status of acquisition programs and communicating about it are in everyone's best interest. Again, work done in the planning phase can mitigate some programming and budgeting instability.

It is worth noting that while the CNO has significant influence in requirements development and the POM, until 2016, this position had no statutory authority over acquisition. This sometimes led to institutional interests at odds. While these conflicts are, to a degree, inevitable, resolving rather than elevating them is generally the preferred course of action. The FY 2016 NDAA contained language shifting some acquisition oversight and decision authority from the Under Secretary of Defense for Acquisition, Technology, and Logistics (AT&L) to the Service Chiefs. At the time this guide was written, the implementation of this issue was still unresolved. But, if approved, the implications could be profound because it could impose a potentially significant set of new, acquisition-related functions onto the Service headquarters staffs to support their chiefs' new authorities.

Other Related Processes

U.S. Marine Corps PPBE Process
The USMC submits a POM just as the Navy does, and both are combined into a balanced DoN POM that is integrated by the ASN's Financial Management and Comptroller (FM&C) office and submitted to OSD. Despite numerous interdependencies between Navy and Marine Corps programs and budgets, there is an occasional tendency to treat the two as completely separate until they reach the Secretariat level. This may result in unpleasant program shocks that could probably be avoided by better and earlier understanding of challenges and coordination between Service counterparts.

More specifically, while USMC has some acquisition programs, its aviation is in Navy appropriations, its amphibious lift is entirely funded and maintained by the Navy, and big chunks of sustainment are provided by the Navy. These are important parts of the national force posture and must be supported, even if some in the Navy do not agree with the USMC share of the combined DoN POM or budget. Conversely, not every USMC requirement is critical, let alone priority enough to override Navy equity. Early informed discussions should lead to better decisions.

Note that the battle over resources rarely takes the form of an argument over total obligation authority shares. It comes instead when sponsors or programmers in need of offsets make cuts on programs that USMC views as critical within the Navy Program (for example, amphibious shipping or connectors). Nevertheless, because the Secretary of the Navy (SECNAV) must balance the two Service POMs before submitting the DoN POM to OSD, there is always potential for some tension that does not come up in the Army or Air Force. Also, the two DoN Service POMs must deliver on a timeline that is three months shorter than their Army or Air Force counterparts because the SECNAV and his or her staff must have time to review and reconcile the POMs into a coherent department submission.

U.S. Department of Defense PPBE Process
As stated earlier, PPBE is a DoD-wide process in which the Navy is one participant, along with the other Services and defense agencies. Navy's POM calendar mirrors that of OSD and reflects guidance that OSD provides early in the process, such as fiscal and programming guidance. At a higher level, OSD provides the national-level Guidance for Development of the Force, with force development following prescribed defense scenarios. CAPE's role in planning and programming in OSD is similar to that of OPNAV N80 and N81, and it is the primary review authority for department POMs. OUSD (Comptroller) has similar roles to ASN (FMB) in budgeting and execution.

As mentioned earlier, OSD operates an issue team process that, for most of the 2000s, resulted in a combined program budget review. Services would submit POMs, and then OSD and other stakeholders would assess these for compliance with guidance and for effectiveness in meeting identified gaps. This budget review has been held concurrently with the review of specific budget actions, most of which are technical in nature. However, the Deputy Secretary of Defense stated that the program and budget reviews would be done sequentially in the POM 2017 cycle. Although the details of this process were still being developed while this document was being written, historically, the sequential process involved significant interaction between the predecessor to CAPE (the Office of Program Analysis and Evaluation) and Navy programmers through late summer. OUSD (Comptroller) and ASN (FMB) counterparts then had more time to work through details, and potentially affect programs, more significantly than they currently might under the combined process. Whatever shape the process takes, however, a continued close relationship between N80 and FMB counterparts remains essential to a successful POM submission.

Summary

This chapter has described very complicated processes in only the broadest terms. There is a lot to know about all three DoD decision-support processes, and some leaders spend most of their careers learning about just one. But one can be effective sooner and more consistently by remembering and understanding that the processes interact with, depend on, and influence each other. It is not possible to treat one decision in one process as something made in isolation. A requirements decision triggers an acquisition decision, which creates a demand for resources. Curtailing resources affects acquisition and may affect the ability to meet a requirement. Moreover, numerous organizations have equities and direct participation in PPBE. Failing to inform and coordinate can lead to surprises and extra work. The Navy's PPBE process is an input into the DoN's PPBE process, whose output is an input into DoD's PPBE process.

A key theme that readers will see reinforced throughout this guide is the collective need for establishing and maintaining relationships and communications with a range of PPBE stakeholders throughout the process. Individual and collective success depends on such interactions, so it is worth the considerable effort to maintain them. Who the key stakeholders are for a particular job will come with experience (see Chapter Four for more information on this topic). The next chapter describes important details of how Navy staff perform the PPBE process, with particular emphasis on the parts of the process that are more or less enduring, including descriptions of key players, timelines, and repetitive processes. The OPNAV organization has changed repeatedly, but most of the steps in the PPBE process have remained essentially unchanged since the system was first implemented.

CHAPTER THREE

How the OPNAV PPBE Process Works

This chapter begins by describing the key functional roles in the Navy's PPBE process and who fills those roles. It then describes the process and its timeline. It concludes by highlighting the key phases of the process, as well as important decision points and deliverables.

Key Functional Roles

Although the roles and relationships among specific OPNAV stakeholders in the PPBE process are addressed in more detail in Chapter Four, some key stakeholder roles are of central importance to understanding OPNAV's PPBE process. Their roles form the foundation for

- recurring supporting and supported relationships that must be understood if one is to be a productive participant
- the types of communications that need to occur between stakeholders at each step
- how those roles and communications change as the POM winds its way through the PPBE phases and up the chain of command.

Some roles are explicit (defined by statute, policy, or directive) while others are implicit (derived from practice or encountered in various POM-related materials, such as POM serials). We use roles in this guide as a way to help the reader understand the types of responsibilities and activities for which stakeholders are accountable in OPNAV's PPBE process.

Capabilities and Resources Integrator (N8)

Navy's PPBE process and organization rest on the foundational notion that resources, requirements, and capabilities require integration, which is a responsibility filled by the N8. The N8's role has varied over the years, particularly in relation to the platform sponsors (see below), but the role as principal executor and developer of key parts in the

POM and budget processes has not changed. The N8's traditional functions include the following:

- orchestrating and integrating the Navy POM
- developing and coordinating the Navy's annual POM and Program Review strategy
- developing and coordinating the Navy's annual POM narrative and communicating it to external stakeholders
- assessing, validating, and integrating Navy capability, program, and resource requirements
- balancing the Navy's requirements and resources portfolio
- balancing the Navy Program
- delivering and defending the Navy POM to the SECNAV
- defining OPNAVs POM schedule for coordinating the major deliverables and deadlines and keeping the process on track
- publishing and enforcing the POM guidance that defines the "rule set" for each phase of the Navy PPBE process
- having the final say on the Navy POM, or tentative POM, before it reaches CNO review.

The specific authorities associated with each of these functions varies by who is CNO, how he or she chooses to do business, and, often, by when the POM occurs in the CNO's tenure. Nonetheless, the N8's position always involves some type of integration and adjudication of competing demands. At times, the N8 has not only had responsibility for the POM and associated processes but also for development of SPPs. In 2011, a new CNO opted to separate resource sponsorship and the N8's role. Some CNOs are exclusive in how they interact with the N8, making planning and programming decisions in one-on-one interactions with the DCNO. Others are more inclusive and involve other DCNOs in their deliberations. This has changed before and may change again, but the N8's integrating and balancing function will likely remain.

Programmer (N80)

N80 is responsible to the CNO and the N8 for all matters regarding the development of the Navy's POM and for all technical and procedural matters concerning the PPBE process and the JCIDS. Acting in this capacity, N80 leads the development and defense of the Navy Program and coordinates the generation, maintenance, and review of all Navy programmatic documentation. The division is also responsible for overseeing the Navy Capabilities Board and R3B processes. N80 consists of three branches— Program Planning and Development (N801); Integration and Alignment (N802); and Navy, Joint, and Urgent Requirements (N803).

N801 is principally organized to develop and deliver the Navy POM on behalf of the CNO. It establishes fiscal and programmatic guidance, develops the Navy's POM strategy, and authors and publishes a series of guidance documents called *serials*. This branch integrates the SPPs into a comprehensive Navy Program, develops the final POM balance for leadership consideration, and ensures program and budget integration with ASN (FMB) or OSD for final recommendations for the Budget Estimate Submission or President's Budget. It also manages the Program Budget Information System in partnership with ASN (FMB).

N802 facilitates, coordinates, and enables internal and external division functions and activities. In those roles, it provides inputs to the CNO's annual posture statement for congressional testimony, supports the N8 for testimony and hearings, and supports the DoD's Defense Management Advisory Group and Three-Star Programmers committee (all Services) by issuing coordination briefs. It also develops external messaging for the CNO, N8, and N80.

N803 manages the Naval Capabilities Board and R3B processes. It serves as liaison between the OPNAV resource sponsors and OSD, the Joint Staff, and combatant commanders on the process for developing Navy capability documents. It performs JCIDS gatekeeper functions for the Navy, supports Navy representation to the Joint Capabilities Board and the JROC, and handles urgent needs.

Resource Sponsor

The Navy PPBE process uses advocacy to deliver a balanced POM that addresses the range of warfighting requirements. At times, this advocacy has reflected Navy leadership's belief in "competition" or "healthy tension." At other times, the use of advocacy has reflected the belief that getting a complete story depends on getting many different perspectives. Resource sponsors are the Navy's capability, program, and requirements advocates. Those sponsors include the DCNOs for Manpower, Personnel, Training, and Education (N1); Integration Dominance (N2/N6); Fleet Readiness and Logistics (N4); and Warfare Systems (N9).

Resource sponsors have varied between being three-star "platform barons" with more or less complete authority over platform-based programs to two-stars under the N8 and, most recently, within an N9 organization. But the fundamental duties have not changed significantly. Responsibilities have typically included

- representing Fleet requirements associated with the sponsor's capability portfolio
- defining, advocating, and defending the requirements in the sponsor's capability portfolio
- defining and translating the capability requirements portfolio into program and resource requirements
- vertically integrating the sponsor's subsets of the Navy's program portfolio into complete capabilities

- developing, submitting, and defending SPPs for the sponsor's program portfolio to the N8 and the CNO
- integrating capabilities and resources for the program portfolio
- integrating the aspirations of applicable PEOs, PMs, and SYSCOMs into the sponsor's internal SPP and eventually combined SPP; this includes statements of requirements on such topics as maintenance and modernization and other support elements.

The most obvious resource sponsors are those associated with the Expeditionary Warfare (N95), Surface Warfare (N96), Undersea Warfare (N97), and Air Warfare (N98) platforms, often referred to as the "high-9s" or "platform sponsors." However, the N1, N2/N6, and N4 also function as resource sponsors for programs allocated to them and issues slightly different from N9 issues. Because both the N2/N6 and N4 sponsor major platform programs, they are also sometimes referred to as platform sponsors. N84 (the Chief of Naval Research's OPNAV hat) and N89 (Special Programs) are also resource sponsors.

In the summer of 2015, an additional resource or platform sponsor—N99 (Unmanned Warfare Systems)—with narrowly defined portfolio authorities in unmanned systems and experimental and prototyping capability innovation was established within N9 in response to a SECNAV initiative.

The scope of each sponsor's portfolio is decided by the CNO, and programs move from one sponsor to another from time to time. Usually, these moves are associated with shifts in responsibility associated with staff organizational realignments, but sometimes, they are the result of leadership's shifts in priority or emphasis. Where a program is situated within the staff will typically affect the types of relationships and conversations that become part of the routine planning and programming for that program, including how capability and resource priorities and trade-offs are framed and decided.

Requirements Sponsor

Resource sponsors deal with particular aspects of the Navy's overall requirements but might not have responsibility for identifying the Navy's "total requirement" for such matters as manpower, maintenance, intelligence, operations, and a variety of other areas that cut across programs and for which there might not be a natural advocate. Requirements sponsors are responsible for highlighting issues of collective importance and that usually require collective or centralized actions to mitigate or resolve. Examples include the OPNAV N1 for overall manning levels; the N2/N6 for ensuring interoperability; the N4 for maintenance and readiness; and the N8 for force structure and total force capability and capacity (N81), ordnance requirements (N81), and special programs (N89). In short, a requirements sponsor is responsible for advocating and/or assessing a set of requirements, although the sponsor does not necessarily own the resources for all of those requirements during the PPBE process.

For example, Navy leadership pays close attention to the N1's assessments and evaluations involving manpower accounts because corporate or collective actions are usually required (that is, no single resource sponsor can resolve the accounts on their own). This is similarly true in other areas, such as spares, maintenance, training, and ordnance.

Process and Timeline

We introduced a generic timeline in Chapter Two (Figure 2.3) that showed the overall PPBE process, all the way through congressional budgeting and execution. The time-line in Figure 3.1 shows greater detail, including some of the timing overlaps between the DoD and DoN PPBE processes, as well as the overlaps within OPNAV's process.

Through the late 1990s, the normal process was for full POMs to be developed only every other year, with more-limited program reviews completed in the off years. Now, a POM is developed each year. Other than this change, the components and timing of the cycle have remained essentially constant. Through wars and crises, the PPBE process has continued and POMs have been delivered. An annual "POM cycle" (the DoD or Navy programming phase) typically begins in September and is normally kicked off when N80 publishes POM Serial 1, which lays out the POM schedule, deliverables, and stakeholder duty assignments. Figure 3.1 illustrates three successive POM cycles and shows how, at any given point in time, three different cycles are in play—each in a different phase of the DoD's PPBE process. This also underscores that, often, AOs and decisionmakers must be accounting for and responding to issues related to different fiscal years simultaneously.

Key Phases, Decision Points, and Deliverables

Figure 3.2 adds key stakeholder processes (in red) to the schedule introduced in Figure 3.1. Notice that processes regularly overlap, which tends to compress activity timelines at key points in the process. Sponsors, for example, must begin preparing program proposals for inclusion in the POM even as the assessment phase is in prog-ress. There have been many attempts to smooth these overlaps without much success. The bottom line is that they tend to occur at known times in the year. Therefore, prior preparation, coordination, and clear understanding of the CNO's and a particular chain of command's intentions are critical to avoiding mismatches and wasted time.

POM development does not occur in isolation from the previous year's submis-sion. Previous cycles will have set resource levels or deferred some decisions requiring action in the current cycle. Moreover, trade-offs or recommendations in one phase of the cycle may be operating under a different rule set than previous phases. For exam-

Figure 3.1
Simultaneous Subcycles Within DoD's PPBE Timeline

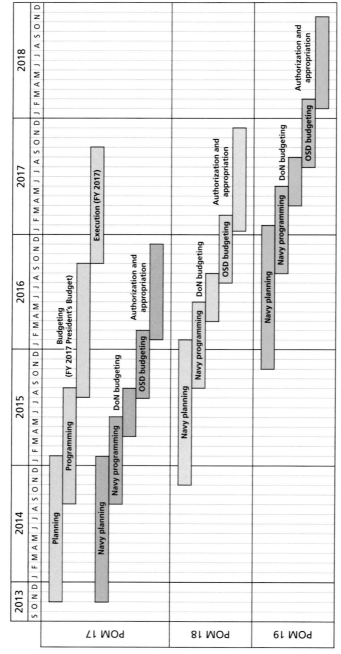

SOURCE: DoN, 2015a, 2015b.
RAND TL224-3.1

Figure 3.2
Stakeholder Processes Within the PPBE Timeline

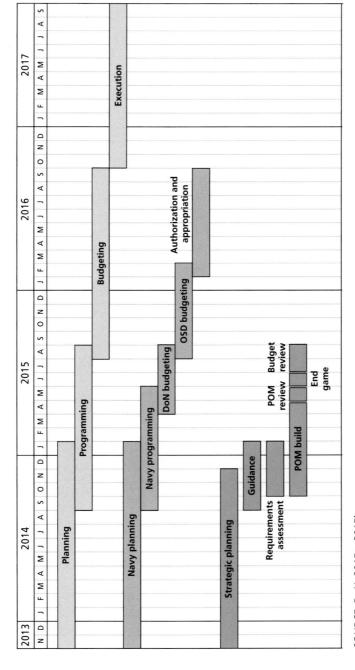

SOURCE: DoN, 2015a, 2015b.

RAND TL224-3.2

ple, a planning phase decision to emphasize one capability over another is, at best, a recommendation and might just be an invitation to think about something. In the programming phase, a decision to deemphasize a capability has greater immediate effect, creating doubt about a program's future. If there is action taken against a program in the budgeting phase, the effect is immediate and potentially significant. The execution phase deals with already appropriated funds, so decisions are less about funding level than about timing and ability to execute.

Strategic Planning

Strategic planning is an ongoing activity that is not necessarily tied to the calendar but is still mindful of it. There may be occasions when some strategic product will be timed to influence a POM cycle, but strategic planning documents often stand outside the PPBE process. With that said, such documents as the Defense Planning Guidance, the NSP, and the SECNAV's *A Cooperative Strategy for 21st Century Seapower* (USMC, DoN, and U.S. Coast Guard, 2015) provide a context and rationale for decisions made at every stage of the PPBE process. These documents are more likely to be influential if they are relatively specific about priority and are not just a statement of worthwhile goals and desires. Even when timing is unfavorable, working drafts of these documents can be used if members have some confidence in the likely final content. Again, coordination with the relevant organizations (at the AO and leader levels) can help leverage content from planning products for the POM even before those products are published.

Generally working through the resource or requirements sponsors, communities also develop plans that bear on development of new platforms, modernization and maintenance of old ones, and career progression for community personnel. These plans generally align with broader strategic perspectives but exist primarily to articulate community concerns and priorities, as well as to make the community's case to broader audiences. For example, each warfare systems sponsor (N95, N96, N97, and N98) typically has an annual or long-range plan coordinated with its associated Type Commanders (TYCOMs) and warfare area communities. Also, during FYs 2014–2015, N9 pursued a Fleet architecture whose purpose was to cement seams across all the warfare and support domains. These plans eventually contributed to a strategic communications "message" or "narrative," expressed in briefs to the military leadership, OSD, Congress, and other external stakeholders, that could influence the themes and priorities in an SPP.

The Navy also pursues an analytic agenda, coordinated by N81 with inputs from other OPNAV codes, that attempts to add a degree of rigor to longer-term capability and capacity decisions by funding or sponsoring studies, analyses, modeling, or assessments on key topics of interest. Results from this agenda are intended to highlight the relative value, consequences, and risks associated with current and future operational, programmatic, and resources decisions. These results can influence decisions in the

FYDP, but they also give additional definition to capabilities and priorities that the broader strategy might impose.

The analytic agenda is further intended to assist in answering questions for which there may be no obvious programmatic answer. Assessments performed in previous cycles may influence the agenda, or they may be an attempt to influence upcoming cycles. Items may be included as the result of resource or requirements sponsor input or of direct interest by senior leadership. Ideally, these items can be dealt with in the cycle imposed by the PPBE process to ensure that answers with a programmatic implication can be appropriately addressed.

Guidance

Guidance comes in many forms, from very broad strategic guidance to the more detailed guidance of POM serials. In general, strategic guidance results from strategic planning and does not follow any particular timeline. However, outright failure to comply with guidance is a clear route to a rewritten or rebalanced program.

Closer to the immediate environment of OPNAV, it is worth reiterating that the Navy POM submission is the CNO's document and that his or her executive agent for developing it is the N8, particularly N80. Early in the POM cycle, N80 issues serials that give sponsors the outlines for POM schedules, duty assignments, deliverables, and deadlines that should be considered foundational for POM development. The serials may provide some latitude for sponsors to prioritize one capability over another, but they provide enough information about priorities that sponsors should understand the limits of choice. Responding to this guidance is another case in which communication at every level is likely to result in an improved product. Guidance would not be promulgated as guidance if it were intended to be close hold.

Worth noting is that POM serials with attendant priorities are generally issued prior to completion of the requirements assessment process (described in the next section). N80 is in charge of the serials, and they provide explicit programming guidance as approved by the CNO. The N8 will normally act to enforce that guidance, and it is understood that it comes from the CNO. It is possible that mismatches between initial guidance and assessment recommendations might occur. However, in general, POM serials are issued with some sense of where the assessment might be trending and with sufficient latitude to incorporate findings into sponsor proposals.

Guidance also usually comes from OSD (usually via the Deputy Secretary of Defense) and from the SECNAV as annual memoranda. In recent years, these guidance sources have become more prescriptive and specific, which leads to another challenge that PPBE practitioners must account for in their planning. When the themes and priorities in OSD, SECNAV, and CNO guidance line up, that makes life much easier for sponsors, N8, and the CNO. Everyone generally gets in line and builds their POM in compliance with guidance, and the issues then become at least a little more routine. But when, for example, OSD and SECNAV guidance do not align, that poses significant challenges for the CNO and the sponsors. Whose guidance should you align with, and how should you interpret it? The answers are not as simple as "align with your next immediate superior in the chain of command," and answers are made less simple by uncertainty in the fiscal and political environments and by the billions of dollars that typically constitute each sponsor's program portfolio. These are some of the most difficult decisions that leaders must make in the PPBE process because of the potential short- and long-term consequences to both programs and organizations. Furthermore, guidance decisions typically drive how the POM is put together and how it is justified, which affects the workload and work priorities of the entire affected organizations.

The key takeaway here is to pay attention to guidance. The Navy POM is the CNO's product, and sponsors' submissions will be "graded" and balanced during programming end game in alignment with CNO guidance.

Requirements Assessment

Assessments take place at various levels almost continuously throughout the year. These assessments include the major studies and analyses that are part of the analytic agenda and major POM deliverables, as well as program reviews and quick-turn evaluations of requirements and performance data. OPNAV has processes that formalize program assessments and attempt to influence POM development directly, although, as noted, these do overlap with the POM build. These processes include the Director of Naval Intelligence's Intelligence Update, N81's Front End Assessment (FEA), the N4's Integrated Readiness Assessment, N81's Warfighting and Support Capability Assessment (WSCA) and Integrated Program Assessment delivered at the beginning and end of POM development, and various BAMs. All are intended to capture and price Navy-wide requirements to inform the POM build. These assessments function as a conscience-call for programmers and capability developers. Especially notable is that N81's products are delivered to the CNO and frequently result in directive guidance to the N8 and N80. Also, the results of these assessments frequently make their way into other high-level POM-related presentations to the SECNAV, the Joint Staff, and OSD to make a case for specific priorities or trade-offs. The final product from this process is an update to the next year's analytic agenda.

Fleet Forces Command and the Pacific Fleet, particularly the N8 divisions in these organizations, are important to the assessment process, especially the N81 products and the BAMs. The Fleet tends to have an interest in nearer-term readiness and the availability for current operations. BAMs might therefore occupy more of their attention than assessments dealing with longer-term future capabilities. Moreover, to the extent that recommended future capabilities may affect the ability to fund current readiness, Fleets may view assessments as distractions. On the other hand, Fleet stakeholders validate the future concepts of employment and concepts of operations used in all of N81's major future capability and campaign analysis products. Also, the results of Fleet exercises, battle experiments, and war games influence these same concept plans.

While the BAMs hold great importance for those who authored them, they are considered advisory to the programmers in N80. Many times, the BAMs suggest a set of requirements that are not affordable, and the CNO must decide where to take risk in these areas. In other cases, BAM authors point out errors. For example, years ago, an aircraft carrier was entering the later POM years without a crew, and BAM authors were able to identify the problem. BAM authors should stay in close touch with N80 during the requirements assessment process. Through their inputs to the PPBE process, BAMs directly affect SYSCOMs and the Fleet maintenance organizations. Generally, a BAM is most useful when it is able to show which requirements have changed, and why, from one year to the next. These trend analyses are prerequisites to making a strong case for more than marginal changes in resource allocation.

POM Build

Although the Navy POM is the CNO's product, developed and synthesized by N80 and submitted by the N8, the raw materials for the POM come from SPPs submitted by resource sponsors. Development of these products overlaps with the assessment and planning process, so sponsors are expected to reflect the results of these assessments even as they are putting their POM proposals together. Close attention to CNO guidance is particularly critical in developing proposals that both protect sponsor equities and advance the overall needs of the Navy.

A POM is a resource database that translates requirements of various types into resource proposals that capability developers and Fleets can use for planning and program execution. It is a political statement, in some respects, because it communicates a set of priorities from the CNO, emphasizing some capabilities or communities above others. It is also a benchmark from which the CNO can demonstrate commitment to items before Congress or OSD. SPPs are the first overt step in the process of programming resources to requirements.

A POM deals with the continuing tension of the vertical integration between platforms and their supporting element (e.g., manning, training, maintenance) and the horizontal integration between all the things that result in a Fleet capability at sea. Platform sponsors are looking across a period of years and trying to ensure that resources are in place so that when a capability is delivered, the personnel, maintenance, infrastructure, and training needed for the platform are also in place and sustained through its life cycle. This is a different perspective from a Fleet looking to deploy a current force composed of many platforms that operate together now. A related tension is that requirements sponsors—looking at the total requirement for personnel or maintenance across programs and across the Fleet—have a different perspective from a resource sponsor, whose primary concern is a particular set of platforms. What might be best for the Navy as a whole may not be optimal for a particular platform or community.

Advocacy is inherent in this process and critical for ensuring that a wide variety of views are considered. The Navy is a large and complex organization. No one analytic approach is going to provide a single set of optimal answers. So, despite the friction and hours involved, a competitive process encouraging stakeholders to develop and defend arguments for capabilities helps ensure that decisionmakers understand the issues before making (or deferring) their decisions. Nevertheless, advocacy cannot translate into unwavering zeal. An advocate's perspective is, by design, more narrow than the integrators' and decisionmakers' perspectives, and advocates need to always be prepared to make their points in terms of the contribution to the Navy's overall capability, not just one community's more narrowly defined interest. Arguments focusing solely on potential community harm might be compelling to the community, but they may just irritate those trying to achieve balance across all of the Navy's interests.

N81 provides both (1) a front-end assessment of naval capabilities represented in the program of record against a set of fiscal constraints and (2) a "grading" of the POM in support of that assessment after the SPPs are delivered and adjudicated by N80. Other processes, such as BAMs (for military personnel, maintenance, spares, munitions, and so on), also play a role in the final state of the POM as it heads to the CNO and eventually the SECNAV. Resource sponsors at the two- and three-star levels have a say in the state of funding for their programs. ASN (RDA) weighs in on the industrial base and the status of various major defense acquisition programs, and others in the Secretariat are concerned with the focus area and support programs in their spheres of influence and responsibility.

The Navy POM ultimately is an input to the DoN POM. Besides the importance of aligning with the USMC—or at least communicating areas of common interest— staying aligned with the SECNAV's acquisition priorities is of particular importance. For example, if the SECNAV is prioritizing procurement of some number of ships or airplanes and the Navy POM submission does not support this, the Navy POM will likely be adjusted after submission to ASN (FMB).

Always keep in mind that there are different stakeholders and decisionmakers who adjust the POM on its way through the process—for example, sponsors, the CNO, the SECNAV, the N8, and representatives from ASN (FMB), ASN (FM&C), CAPE, and OUSD (Comptroller). In most cases, each stakeholder is motivated by priorities and rules that are different from their counterparts, and each would likely make different choices. The degree of influence that these stakeholders have on the POM is also unequal and changes as the POM moves from one phase to the next. In other words, there are windows of time during which a stakeholder has more influence over priorities and decisions. Outside those windows, influence may be limited or nonexistent, but the windows are almost entirely predictable because PPBE is a calendar-driven process. This means that stakeholders need to plan and prepare for their windows of influence with their best efforts and supporting materials, and those efforts must be founded on realistic expectations in the context of leadership's priorities and guidance for Navy's program as a whole.

POM Integration

N80 and N81 review SPP submissions to assess the degree of compliance with guidance. After these proposals are presented and reviewed, N80 generates a tentative POM that incorporates guidance, sponsor proposals, and the need for a program balanced across the FYDP. Sponsors are expected to participate in this process, and it is usually the last opportunity for them to influence the overall balance. During this time, the requirements sponsors in the resource sponsors' organizations are, or should be, in direct contact with their N80 and N81 analyst counterparts. In some cases, the resource sponsors may use analytic agenda results to defend their program proposals. There should be a large amount of cross-talk during this period.

End Game

As the tentative POM is reviewed and refined, the number of actors who can affect the POM balance shrinks. In the end, the POM must balance in alignment with CNO guidance, and in this process, trade-offs are made and risks accepted, sometimes in places of deep interest to communities and sponsors. As these become known, the next likely steps in the planning process are to identify the potential consequences and propose mitigations. Taking issues outside the system to OSD and the SECNAV may be counterproductive. Sponsors should be prepared throughout the review process to answer what-if questions quickly and accurately. This process is called the *end game.*

Sometimes during the end game, the SECNAV may force a change in the CNO's priorities, either because of external influence (such as a reduction in the topline or a changed national priority) or because of issues in the POM that were not addressed to the Secretary's satisfaction. When these occur, changes in the POM may result, possibly accompanied by rapid turnaround requests for information and impact. During such drills, it remains important to be flexible, responsive, and accurate, even if the possible results are contrary to preference.

N80 brings the POM into fiscal balance during this period, and its analysts are busy, harried, and stressed. They are getting help from (or being distracted by) many sources, including the SECNAV's budget analysts, who will review the POM and turn it into a budget in the final analysis. If phone calls go unanswered and emails unread, consider a face-to-face meeting with N80 staff to determine the state of a particular program(s).

Budget Review

BSOs are offices and major commands that are given resources by the CNO for the program and support activities under their purview. For example, they may receive resources to maintain current-year end strength. However, budgets are subject to constraints and new demands, as are programs, and therefore require adjustment. Budget review is intended to assist in balancing in areas where balance is not just desired but legally required.

As mentioned in Chapter Two, for several years, program and budget reviews have been held at the same time, with issues spilling into each process. In 2015, OSD proposed that these processes once again be carried out sequentially rather than concurrently. This may allow a better separation between short- and medium-term issues and a fuller appreciation of longer-term sustainment requirements. However, it may also compress the time available for programming and possibly make budget review more program-focused.

CHAPTER FOUR

Stakeholders, Roles, and Relationships

This chapter describes the key stakeholders in the PPBE process, both inside and outside the OPNAV staff, that influence and sometimes govern the way issues are raised and resolved. Many of these stakeholder roles and responsibilities may appear opaque to those involved in only pieces of the process, but an understanding of the overall stresses on the system is important to determining how an individual, his or her boss, and his or her organization fit into the process. This chapter is, in many ways, the most important in the guide because without a good understanding of who does what, why they do it, and how they relate to your own responsibilities in PPBE, much of the rest of this guide is of limited use. Your understanding and utilization of these topics form the foundation for the rest of your POM-related activities, affect how you do business, and determine your and your organization's effectiveness.

Key Department of the Navy Stakeholders

Chief of Naval Operations

From the viewpoint of OPNAV staff, only one stakeholder demands attention, and that is the CNO. Everything in the Navy's programming process is geared to the CNO's vision, the decisions that he or she thinks are important, and how he or she wants to do business. That does not mean that the CNO is not beset by outside pressures, and this chapter walks through some of them. Whether you are a resource sponsor, a requirements sponsor, or a programmer or analyst in N80 or N81, always remember that OPNAV's POM is developed for the CNO.

This focus raises the question, How does one know what the CNO desires in the POM? The first and best way is to get feedback from members and leaders attending the CNO's information and decision forums (when this information is releasable). Some CNOs are explicit and will encourage dissemination of key points from these meetings. Others employ a more guarded style. Either way, there likely will be the need to derive intentions by paying attention to the CNO's external communications. These include testimony, leadership conference presentations, public speeches, strategic guidance, and, in the case of a new CNO, the material prepared for the confirmation hearing. The CNO's all-Navy ("ALNAV") announcements, sailing and navigation instructions, similar policy and intention documents, and transmittals are all part of the leader's message and should be considered in POM planning.

Remember also that DoN has two Service Chiefs (the CNO and the Commandant of the Marine Corps) submitting inputs to the DoN POM. Therefore, although the CNO *proposes* the Navy POM, the SECNAV *decides* the DoN POM. The degree of SECNAV involvement has varied by individual leader. In times past, the SECNAV has been fully engaged in detailed briefs on every element of the POM cycle. A SECNAV once decided to deliver the POM almost three weeks late to enable him to make decisions at almost the line-item level. Other Secretaries have stayed away from such detailed management and exercised influence through their guidance and relevant executive decision forums. The methods are less important than recognizing that the SECNAV is a decision authority if he or she decides to exercise it, and the CNO's relationship with the Secretary usually has a significant influence on the guidance (from both the CNO and the SECNAV) and the timeline in which OPNAV must build its POM.

In addition, the Commandant of the Marine Corps plays a role in CNO decisions about those capabilities in which the Navy and USMC are team players—mainly aviation and amphibious ships. In fact, Congress recently legislated that the N95 Director be chosen from the Marine Corps two-star list. While the negotiations between the two Service Chiefs are not public, it is clear that the CNO is influenced by the need to fund and support Marine Corps programs.

For those working for resource or requirements sponsors, it is important to understand that N8 and N80 Directors and their Deputy Directors are in relatively constant contact with the CNO as the POM cycle progresses (depending on how the CNO chooses to do business with the DCNOs). If your position is within N80 or N81, remember that what is heard from the CNO may be privileged and, therefore, not for further dissemination. These discussions are not public, nor should they be, but the N8 and his or her Senior Executive Service members and flag officers speak for the CNO and pay a great deal of attention to and derive guidance from the CNO. A key point here is that all else being equal, leaders' and action officers' relationships with their counterparts in the N8 and N80 offices are usually very important to staying informed by and aligned with the CNO's priorities and intentions.

Fleet Commanders and Type Commanders

Fleet Commanders (FLTCOMs) and TYCOMs have an independent viewpoint and voice in POM issues, and CNOs historically have paid a great deal of attention to their input. Following the TYCOM consolidation in the early 2000s, the U.S. Fleet Forces Command's N8 took an integrating role in TYCOM requirements, with significant regular coordination between the TYCOM and FLTCOM N8s. Fiscal reality, personalities, and near-term and far-term perspectives all come into play in these relationships. Many of the FLTCOM issues that occur are reflected in the assessments developed by N81 and drive many POM priorities and decisions. The Fleets can participate in as much of the POM as they desire, so expect some recurring "battle rhythm" of video teleconferences during certain periods of the planning and programming phases for the POM. Critical Fleet requirements, usually in the form of personnel, training, or maintenance issues, are rarely ignored. FLTCOMs and TYCOMs are routinely in touch with resource or requirements sponsors, as well as the heads of the Navy's Enterprises. Additionally, the N8 and N80 often feel pressure from those organizations.

Sponsor and N8 relationships with their community or counterpart representatives in the Fleet are important to both the process and outcomes of the POM. A steady battle rhythm throughout the year to maintain situational awareness on both sides, augmented with more-focused interactions early and near key deliverables in the planning and programming phases, should be considered part of the normal business rules. These should also correspond with the windows of opportunity most relevant to incorporating Fleet inputs (i.e., in time to be relevant to the subject product or decision).

Systems Commands

As stated earlier, SYSCOMs hold the keys to the technical excellence and authority of the Navy and have expertise in engineering and in acquisition life-cycle management and maintenance. During the POM cycle, they deal with the Fleets, the PEOs, and ASN (RDA), but it is important for resource sponsors to keep apprised of the tough issues that perennially arrive at the doorstep of the POM. SYSCOMs have the ability to provide reliable cost, design, and engineering estimates on a range of topics related to material and software systems. They also have cross-program and historical perspectives that can be very useful in assessing the viability of various options that typically arise when developing and assessing requirements.

OPNAV N8

The N8's relationship with the CNO is typically close, but it depends on how the CNO chooses to do business and involve the DCNOs in POM planning and decisionmaking. The N8 is often regarded as the "first among equals" in the POM process. Under the staff configuration in 2015, the N8 is not a resource sponsor (i.e., owns nothing) and is therefore not a proponent of any single issue. Therefore, the N8 can represent the CNO's interests without bias toward specific requirements or communities. The N8 is typically a direct liaison with the Navy Secretariat, including the ASNs for research and development, financial management, manpower and reserve affairs, and energy and environment, and is charged with reducing or eliminating surprises for the SECNAV. The N8 is a member of the Three-Star Programmers group with CAPE and the other military department programmers. This DCNO also deals directly (through N80) with ASN (FMB) and the budget process, but because that process is a DoN one, the N8 represents the CNO in the same way that the Headquarters Marine Corps (HQMC) Programs and Resources Division (P&R) represents the Commandant of the Marine Corps.[1] The N8 also represents the CNO with ASN (RDA) and the Under Secretary of Defense (AT&L) in various meetings, as appropriate. The N8 supports the Vice Chief of Naval Operations in his or her JROC duties through N80 and is the counterpart of the J8 in the Joint Chiefs of Staff organizational structure.

All of these roles and representation duties are important for AOs and leaders to keep in mind during PPBE. Under the current (as of this writing) staff and fiscal environment, the N8's perspective is dominated by issues related to the integration, sustainment, and balance of programs, capabilities, and capacities. This puts a premium on the effectiveness of the DCNO relationships; because the platform sponsors are aligned under the N9, the N8 is no longer responsible for developing the longer-term capability-to-programmatic vision and narrative. The N1, N2/N6, N4, and N9 must develop a narrative in collaboration with the N8, and the DCNOs need to work with the N8 on how programs are represented in key information and decision forums.

OPNAV N80

N80 is the execution arm for the Navy POM and is sensitive to the CNO's utterances, desires, and guidance, which are largely not public and not written. N80 issues POM guidance (as POM serials) designed to deliver a Navy POM input to OSD on time, within the fiscal controls provided, and balanced in alignment with guidance. N80 also briefs and defends the POM to OSD and in the Navy budget deliberations, as necessary.

The AOs in N80 are normally aware of the latest CNO guidance (through the N8, his or her deputy, or N80's Director). Any decisions to ignore the requests for information should be done with full cognizance of the potential consequences. N80 mirrors the OPNAV staff in that there is always at least one individual who "represents" a resource sponsor's portfolio. Generally, N80 AOs are encouraged to provide transparency to resource sponsors at all times. A comment or guidance from an AO is usually the same as coming from N80. Staff from N80, its Deputy Directors, and N801 are also regularly in contact with the N8 and the CNO.

For N80 to be successful requires effective, ongoing sponsor leadership and AO/RO relationships with their counterparts in N80. Regular Director-with-Director and RO-with-bullpen relationships are critical to developing and delivering an effective POM. Surprises, in this context, constitute a staff failure. Agreement is not expected. What is expected is that AOs and leaders understand the respective sponsor (advocate) and integrator (N80) roles and develop an effective partner relationship that regularly communicates to ensure that everyone understands the CNO's position, the guidance in effect, each other's intentions and motivations, and the programmatic impacts of the options in play. Sponsors who neglect their relationship with N80 should expect

---

[1] See discussion later in this chapter about HQMC P&R, which is OPNAV N8's counterpart on the HQMC staff.

to be surprised and to at least partially cede their programming role to N80, who, if necessary, will balance the program in alignment with guidance, with or without sponsor input.

In the final analysis, N80 controls the POM database and makes the final entries according to N8 or CNO direction. Sensitivity to requests, timeliness, schedules, and deliverables should have the highest priorities for resource sponsors, requirements sponsors, and BAM sponsors as the POM process evolves. If a resource or requirements sponsor is late, the N80 AO will proceed with the information immediately available to them. From a sponsor's perspective, pay attention to what the N80 analysts are saying, and keep in touch with your N80 counterpart. N80 must also work to keep the information flow open between themselves and the resource or requirements sponsors. Choices must be made, priorities established, and programs defined. Only through continuing AO-to-AO dialogue is this possible or successful.

OPNAV N81

Beginning in 1998, N81 worked hard to reconstitute and grow its analytic capabilities and capacity. As of 2015, the key results of the analytic agenda and annual capability and capacity assessments are summarized in the FEA—usually presented in November—which is an assessment of the state of risk in the Navy Program since the prior POM submission. This assessment accounts for both programmatic and threat-driven risks and looks horizontally (across programs) and vertically (major programs, top to bottom) at the entire Navy Program. Depending on leadership's response to the FEA, the results are used to inform the WSCA mid-POM—usually in January. Think of the WSCA as N81's recommendations, based on sponsor-articulated or analytically derived requirements, for what changes in risk should be included in the POM submission currently under development. Leadership's reactions to the WSCA may result in changes to guidance (to emphasize or sustain particular investments or to adjust some timelines) or may simply be taken into consideration as leadership prepares to receive the SPPs.

The first key point is that these assessments are based on work that may have begun the year before and may include finished analyses whose results are still considered valid. This highlights how far ahead N81 tries to anticipate issues and decisions that may need to be supported analytically, as well as why it is important to proactively participate in developing OPNAV's analytic agenda. It is also important to note that the WSCA comes out in the middle of sponsors' period for developing their SPPs. This underscores the need to maintain ongoing relationships and engagement with N81, similar to resource sponsors' relationships with N80, because decisionmakers should not be surprised by results in the FEA or WSCA. A surprising result indicates a staff failure to communicate. There are times when N81 is expected to close the door and make an assessment or recommendations directly to leadership without comment or sponsor input, but such cases should be the exception.

N81 is also expected to "grade" SPPs in the Integrated Program Assessment, in which the main criterion is the degree to which resource sponsors followed the N8's guidance in their submissions. When there is strong cooperation between N80 and N81, expect N8 inputs to factor heavily in N80 POM integration activities and drills, and expect N81's findings to make their way into the N8's POM presentations to the CNO, Joint Staff, and OSD. When this relationship is not strong, N80's products will reflect a mostly fiscal and programmatic flavor in their integration and balancing decisions.

The degree of consideration paid to such products as the FEA, WSCA, and Integrated Program Assessment varies by leadership preference and is heavily influenced by how the CNO chooses to do business and make decisions. N81 plays a larger (or smaller) part in the POM based on the background of a CNO and the pressures on the Navy resources.

As mentioned earlier, N81 is also in charge of Navy studies, and resource sponsors should be aware of such studies, their results, and the implications of those efforts for various POM issues. Of course, many N81 study ideas come directly from resource sponsors or SYSCOMs, so that is one way of influencing the process. However, the independent analysis may not always yield the result desired by the sponsor. N81 has historically been expected to act as an honest broker in such analyses and will strive to be faithful to where the data take the conclusions. Because N81 is an independent analytic arm of the CNO, it may differ in perspective from that of a political appointee, three-star, N80, and even the N8. To the extent that it is possible to participate in N81's analytic agenda, organizations should do so.

Because resource sponsors and other actors in the PPBE process have a recurring opportunity to influence the analytic studies that N81 might perform, this involvement can ensure that the questions are framed appropriately, that the studies are conducted with at least an opportunity for stakeholders to view work in progress, and that answers are at least understood if not agreed upon. Stakeholders should also attempt to participate in meetings or working groups as a study moves through to completion. Stakeholders should resist the temptation to block access to data or prevent issues from being considered. Obstruction will not prevent consideration of the issues, and if one source blocks access to data, the data likely will just be obtained from somewhere else.

Another purpose of the analytic agenda is to develop analytic results to prepare leadership for decisions they may have to make in the future. This is one of the main reasons for N81's existence—looking ahead to decisions that leaders may not yet realize are important or for which the value of analysis may not yet be appreciated (i.e., by the time the need for it is recognized, it will be too late to get the work done in time to inform the decision). Therefore, N81 spends time trying to anticipate leadership interest and reviewing past ("finished") analyses to determine whether past results (and the assumptions on which the analyses were based) are still valid to the circumstances under consideration. N81 staff are the keepers of past analytic work and the collective state of analysis against existing and emerging issues and questions.

From N80's perspective, analytic agenda results may be useful in informing POM choices, especially when there are cross-sponsor issues to resolve. For example, a finding that a particular threat is evolving at a faster rate than expected would likely point to one set of capabilities over another, and thus might make a difference in what becomes part of the Navy POM. While the analytic agenda does not purport to be a detailed program guide, it may offer analyses to resolve different approaches. Increasingly (as of 2015), many key issues in the N8's presentation of POM priorities, shortfalls, trade-offs, and likely consequences and risks are cast in the context of N81's analytic results.

OPNAV N82/FMB

N82 is dual-hatted as the ASN (FMB) office and is in charge of the DoN budget. Under its N82 hat, the division works directly for the CNO managing the execution of the Operations and Maintenance, Navy and the Military Personnel, Navy accounts. N82/FMB does not lead budgeting under its OPNAV alignment. This is an important distinction, and the OPNAV staff should recognize that this organization works primarily for the SECNAV, representing the Commandant of the Marine Corps and the CNO. Under its FMB hat, N82/FMB is associated with the budgeting of the entire Navy and Marine Corps and works directly for the SECNAV through the ASN (FM&C) office.

In simple terms, the budget is the output of the first year of the FYDP generated during each POM cycle. But because the budget is also the result of inputs from BSOs, N82/FMB makes changes to the POM's first year. In this context, the relationships among resource sponsors, N80, FMB, and BSOs are central to the smooth transition between the programming and budgeting phases; the state of these relationships also can determine how many POM changes to anticipate during the budgeting phase. It is important to understand that the rule sets that N80 and FMB use to develop and evaluate resource allocations are different, in alignment with their principal functions, which reinforces the need for strong relationships and communication.

N82/FMB has a different perspective than that of other OPNAV staff in that it must pay considerable attention to program *executability*—that is, the ability of a program to actually execute its funding, if appropriated. N82/FMB considers congressional guidance, and if programs have shown a poor execution track record with the funds provided, those funds are moved to other priorities. The executability and *rampology* (the rate of increases or decreases in funds for a program year to year) of program lines are key metrics, so ROs should remain attuned to their program's ability to execute and consider this when deliberating requirements. Also, because the budget is built from inputs from the BSOs, whose perspectives may differ from resource sponsors, do not assume that the N80 analysts and N82/FMB budgeteers are in agreement on various issues. Resource sponsors work with N80 to allocate (i.e., program) and balance resources across their programs based on estimates of required resources over time. BSOs submit their estimates of how to execute programmed resources over time, and their estimates may not align with the sponsors'. Ideally, sponsors and BSOs would agree, but the timing of these estimates during the POM cycle sometimes precludes this, and BSOs and sponsors sometimes simply disagree based on their own resource or requirements priorities. Again, ongoing relationships between these stakeholders and communication about programs' statuses are important to smooth transitions between programming and budgeting.

ASN (FMB) integrates the DoN budget for delivery to ASN (FM&C), with emphasis on how well programs are achieving or can achieve their milestones (these program reviews take place within FMB-2, the procurement and research, development, test, and evaluation branch). For the Military Personnel and the Operations and Maintenance, Navy accounts, a similar series of execution reviews are held (within FMB-1, the operations and support branch). Armed with that information, budget analysts visit the SYSCOMs, PEOs, and Fleet activities to review the status of projects and view the changes made by OPNAV during the PPBE process. Senior leaders in FMB may attend the relevant SPP briefs during the POM build to become knowledgeable on the issues. For Fleet issues and maintenance, the BSOs play a larger role than the resource sponsors because they are closer to what is happening in those arenas.

FMB pays greater attention to the execution of Working Capital Funds and to guidance measures (such as inflation indexes) from OUSD (Comptroller) and the Office of Management and Budget. FMB chairs the Program Budget Coordinating Group, which attempts to adjudicate issues that arise between the POM and budget and between the Fleet and OPNAV. FMB tends to focus on the budget and FYDP years and will usually make changes to out-years only as they are affected by budget-year changes.

Although FMB and N80 occupy different parts of the PPBE process, programming decisions can constrain budget decisions, and budgeting can make POMs little more than wishful thinking. N80 is responding to guidance from the CNO. FMB, while having ties to OPNAV under its N82 hat, fundamentally works for the SECNAV and responds to guidance from the Office of Management and Budget and OSD on specific matters of policy. FMB would likely bring an OPNAV programming discrepancy to the attention of leaders in the SECNAV's office. Similarly, a POM that did not respond to guidance on science and technology expenditures would likely receive similar SECNAV visibility. Programs that cannot execute programmed resources (perhaps because of operational scheduling, system availabilities, or workforce issues) likely will be assessed as at risk, and FMB likely will reduce their resources to executable levels. If a POM does not contain out-year funding for a program, FMB will likely see no reason for a line to continue in a current-year budget. For that matter, budget executors might begin looking for ways to move current-year execution funding into other lines. While balance must be achieved across the Navy budget, the Navy generally benefits when alignment occurs more broadly among the components of the PPBE system. The effectiveness of the relationships between resource sponsors and BSOs, resource sponsors and N80, N80 and N82/FMB, and N82/FMB and BSOs are collectively some of the most important indicators of how much work a POM will be in the transition to balance during programming and end game, during the transition to budgeting, and during budget reviews.

OPNAV Resource Sponsors

Four DCNOs are responsible for planning and programming most of the Navy's resources. This section touches on some noteworthy roles and relationships.

N1

The N1 is a resource sponsor, a requirements sponsor, and a BSO. This is because the N1 (similar to other OPNAV codes) wears multiple hats. This position is both (1) the DCNO for Manpower, Personnel, Training, and Education, an Echelon I staff code reporting to the CNO, and (2) the Chief of Naval Personnel, the Echelon II Commander of the Navy's entire manpower, personnel, training, and education shore establishment. This means that elements of the N1's staff codes are actually Echelon II staff, under his or her Chief of Naval Personnel hat, performing both Echelon I (planning and programming) and Echelon II (BSO, budgeting, and execution) work. Because of these responsibilities, the N1 also has a unique role as holder of the manpower database. Therefore, the N1 works with FMB to ensure that the manpower account is fully funded in the budget (i.e., the Navy has the money to pay its people). The N8 and N80 pay attention to the N1's pre-SPP and post-SPP assessments. If the N1 raises an issue with a resource sponsor, the affected AO should move to adjudicate that before being forced by N80 to pay a bill during the end game.

The N1 is the resource sponsor for all accessions and advanced education, and he or she exercises administrative control over Navy manpower and training policy, reporting, and assessment. The N1 also exercises centralized supervision and coordination of the Navy's manpower, training, and education requirements and advises manpower resource sponsors. This functional alignment is the result the 2011–2012 OPNAV realignment that shifted resource sponsorship authority for, among other things, program-specific manpower and training requirements back to the platform sponsors (the N2/N6, N4, and N9). The N1 is responsible for working with resource sponsors on their platform manpower, manning, and community planning and programming efforts.

The distinction between manpower and manning planning and programming is worth noting. Manpower requirements are related to the determination of the correct number and types of billets (positions) to ensure that the platform, unit, or installation can accomplish its mission (assuming the right types of people are actually assigned and have the appropriate skills and training). Manning is the process of ensuring a sustainable flow of qualified people into and out of those billets. Platform sponsors have to account for both manpower and manning in the requirements and resourcing processes. The N1 does as well, but the key difference is that the N1 oversees the entire database of military manpower requirements and the process of managing the policies (accessions, bonuses, controls, and so on) needed to keep a stable workforce of people moving through those positions. The N1 also delivers the annual Manpower BAM (a report that assesses the resources required to fund everyone's stated manpower requirements) and assesses the overall health of Navy manpower, training, and education for the N8 and CNO. Pay attention to manpower pricing assumptions as they relate to your programs' manpower resource requirements estimates.

N2/N6

The N2/N6 is another dual-hatted resource sponsor, serving as both the DCNO for Information Warfare and as Director of Naval Intelligence. The N2/N6's portfolio includes resource sponsorship of Navy command, control, communications, computer, intelligence, surveillance, and reconnaissance (C4ISR); cyber, information warfare; electronic warfare; Maritime Domain Awareness; and space and Naval oceanography-related systems and programs. Many of the N2/N6's requirements are driven to a major extent by the planning and programming activities of platform sponsors, on whose ships, planes, and submarines many of systems must be installed and maintained. These systems include radios, terminals, antennas, networks, and many other information technologies. C4ISR programs have also risen in importance in every succeeding administration and likely will play an even greater role for CNOs in the future.

A recent change that established OPNAV N99, an unmanned systems and innovation sponsor within the N9's office, resulted in the realignment of sponsorship authority for some unmanned systems programs, shifting from the N2/N6 to N99. N99's role in the POM was still evolving at the time of this writing, but it is expected that N99 will sponsor the innovation and development of selected initiatives and programs, with the intention to migrate them to other platform sponsors in the N2/N6 or N9 at designated points in their development cycle after Milestone A.

The N2/N6 sponsors many of the C4ISR and networking programs on which N9 warfare system effects chains depend. Because the N2/N6 has programs that intersect the resource sponsorship of the N9, integration issues emerge that require regular coordination throughout the POM cycle. Recognize that many issues arise in later stages of POM development, and they typically involve integration and interoperability. The N9, N2/N6, and ROs need to work closely to anticipate and resolve such issues before they reach N80, whose staff have time constraints that make resolution of complex issues more challenging. The N2/N6's requirements and resource integrator is the Warfare Integration Directorate (N2/N6F) and is usually the best place for coordination.

N4

The N4 is both a resource and requirements sponsor whose portfolio includes Fleet readiness, logistics (including Combat Logistics Force and other Military Sealift Command vessels), shore readiness, Navy expeditionary medical services, and environmental programs. Similar to the N1's manpower assessments, the N4 produces both a pre- and post-SPP readiness assessment across the Navy that N80 takes into consideration (including the Readiness BAM). Make certain that you understand the issues that affect your programs and that you either agree with the facts as presented or can refute them. The usual issues involve readiness accounts, such as the ship or aircraft maintenance and operating accounts. The N4 runs Navy readiness models that affect estimates of future requirements and pricing used by N80, N81, and FMB. Working with BSOs, the N4, and N81 early in POM planning is usually needed to avoid surprises and ensure that planning and programming assumptions are consistent between sponsors, the N8, and FMB. Readiness pricing and executability assumptions (including maintenance, spares, and the price of fuel) are frequent sources of disconnects and causes for adjustment during both programming end game and budgeting.

N9

As the sponsor for warfare systems, the N9 is the advocate for the largest proportion of the Navy's total obligation authority each year and is responsible for determining, integrating, and resourcing expeditionary, surface, undersea, and air warfare requirements and their associated systems, manpower, training, and readiness requirements.

AOs and ROs for the N9 should coordinate with their platform and system counterparts in N81, N80, and the N9's Warfare Integration Directorate (N9I). N9I's role as integrator is similar to that of N2/N6F in that it is responsible for pulling together the N9's POM narrative and integrating the platform sponsors' SPPs into a coherent, integrated SPP for the N9.

Because the N9's warfighting capability and capacity requirements drive practically everything else in the Navy Program, how this DCNO interprets and prioritizes leadership's guidance has at least some effect on every other sponsor's POM submission. If you are an AO or RO for the N9, you need to establish and maintain working relationships with counterparts in N80, N81, and the offices of the N1, the N2/N6, and the N4 (as applicable) to ensure that you understand the state of guidance, risk, and interdependence with other resources and requirements issues. If you work outside the N9, one or more programs under that office are likely to dominate your attention and affect your own requirements and resourcing issues throughout the POM cycle.

OPNAV Platform Sponsors

Platform sponsors obviously play major roles within their respective DCNO organizations, and they build (or contribute major portions of) the DCNO's SPP. Platform sponsors have close relationships with platform PEOs and PMs, as well as various Deputy ASNs in the Secretariat. They are more sensitive to changes that may be occurring in the execution of acquisition category (ACAT) programs because of the resource magnitudes usually involved and the attention those programs receive. Platform sponsors include the N2/N6F, the N4's Strategic Mobility and Combat Logistics Division (N42), N95, N96, N97, N98, and N99.

Platform sponsor ROs need to maintain relationships with counterparts in N80, N81, the N1 office, and the N4 office to ensure alignment and to avoid surprises. The quality of relationships between platform sponsor ROs and their bullpen counterparts is usually a good indicator of how the POM is going to go. Most platform ROs outside the N2/N6 office also need to maintain some relationship with counterparts within the N2/N6 office for those systems installed on their platforms.

Each platform sponsor typically has a close relationship with its corresponding Navy TYCOM (e.g., N95 with the Navy Expeditionary Combat Command and Naval Surface Force, N96 with the Naval Surface Force, N97 with the Naval Submarine Force, and N98 with the Naval Air Systems Command). Each of these relationships is somewhat unique, driven by mission and community cultural factors. TYCOMs affect sponsor priorities in important ways because they are the waterfront three-star most directly aligned with and affected by the output of each sponsor's POM efforts. TYCOMs are responsible for preparing and maintaining the platforms and systems advocated and paid for by their sponsors. As a result, there may be some tension between platform sponsors and other leadership (for example, DCNO, N8, CNO, or SECNAV) priorities and guidance. This tension becomes manifested in how the sponsor crafts its SPPs and how closely they are aligned with a specific leader's guidance.

OPNAV N3/N5

Along with its numerous strategy, policy, and operations responsibilities to the CNO, the DNCO for Operations, Plans, and Strategy (N3/N5) produces key planning documents that are more or less foundational to the Navy POM. These include the Maritime Strategy, Naval Operational Concept, Sea Power 21, CNO's Sailing Directions, CNO's Navigation Plan, and the NSP.[2] In general, all of these products are closely aligned and reflect common priorities and themes. While these guidance documents are not normally programmatic in nature, they do set the framework for the POM and are the basis for the CNO's testimony and posture with OSD in the program review.

The N3/N5 may represent the CNO for some or all of the National Defense Strategy, Defense Strategic Guidance, National Military Strategy, Defense Planning Guidance, and various Joint Operating Concepts emanating from the Joint Staff or the OSD staff, as appropriate. To get input into those documents, AOs should be in touch with N3/N5 counterparts. While not a major player in the POM as a resource sponsor, the N3/N5 theoretically sets the strategic framework and has roots in the entire DoD and think tank policy camp. As a result, this DCNO expresses a viewpoint that is often mirrored in higher DoD and executive administration circles and should be considered in POM deliberations. However, in recent POMs, the guiding strategic framework has been the Defense Strategic Guidance, which has been the Navy's primary yardstick for externally discussing its capabilities and capacities.

---

[2] The NSP has also been called the Strategic Planning Guidance. It is normally a classified document.

The NSP is normally the strategy document most closely linked to the POM under development because it includes the *risk guidance*—a set of prescriptive statements that indicate leadership's intentions for reducing, maintaining, or accepting additional risk in the Navy's capabilities and capacities. The specificity of this guidance varies but is normally general enough to accommodate some changes over the course of the POM. During periods of high political and fiscal uncertainty, it can be a challenge to write the guidance. N81 typically has a significant role in drafting the risk guidance, in close coordination with N80 and the N3/N5's Strategy and Policy Division (N51). This is because the risk guidance is usually driven by the CNO's interpretation of the results of N81's mission analyses, campaign analyses, and capability assessments; SECNAV and OSD guidance; and fiscal constraints.

Headquarters Marine Corps Programs and Resources

USMC has its own POM that must go through the SECNAV. It is largely manpower-based but has some acquisition and support programs. The Marine Corps usually has only a few ACAT I programs but has great influence on various naval aircraft and ship programs, as mentioned earlier. USMC's major equities in the Navy POM are found in both shipbuilding and aviation programs. The Marine Corps is also interested in aircraft maintenance and operating resource levels because these affect USMC aircraft and because USMC readiness posture is maintained (by policy) at higher levels than Navy units. Remember that Marine Corps officers also sit alongside Navy ROs in N98 and N95, and a USMC two-star leads N95.

The Navy buys and maintains amphibious ships used by USMC and its aircraft. As a result, these procurement issues are adjudicated in the Navy POM with USMC as a major player. HQMC P&R plays a key role in this process and has direct access to the Commandant of the Marine Corps, and thus to the CNO and SECNAV. HQMC P&R is the N8's USMC counterpart, and counterparts within both organizations have a continuing relationship throughout the POM cycle. Disagreements between Services are both familiar and normal because their responses to fiscal constraints account for different portfolios and postures. Knowing the issues and the history up front is critical and far better than trying to adjudicate them in front of the two Service Chiefs or the SECNAV. The bottom line for AOs and ROs is to scrutinize any requirements or resource issues in your programs that could affect (or be interpreted to affect) "blue-in-support-of-green" positions to ensure that both Navy and USMC counterparts understand each other's intentions.

As an artifact of history, there have been times when the Commandant of the Marine Corps goes directly to the SECNAV and gets topline funding from the Navy. As a separate Service Chief, he or she has the authority to make such a request, although such action is rare. USMC also has direct access to Congress, and USMC leaders are very good at leveraging this access. USMC P&R has a full-time field grade officer (at O-5 rank) in N80 who serves as a liaison between N80 and P&R. Because the Commandant talks with both the SECNAV and the CNO on a regular basis, it is always a good idea to have such a liaison.

Assistant Secretary of the Navy (Research, Development, and Acquisition)

ASN (RDA) is the Navy's chief acquisition executive office and has a major collaborative role in the Navy POM because the ASN has a decision role in every acquisition program. Every platform sponsor must keep his or her counterpart in ASN (RDA) informed so that there are no surprises. Also, remember that the PEOs work for ASN (RDA), not the CNO.

The CNO's role in the acquisition process has been limited to an advisory one since the Goldwater-Nichols Department of Defense Reorganization Act of 1986, but language in the FY 2016 NDAA shifted some acquisition authorities and responsibilities to the Service Chiefs. Exactly how this will be interpreted and implemented and how this language will affect the POM process remained to be seen as this guide was being drafted.

In general, ASN (RDA) does not like programmatic changes in acquisition programs made outside the PEO structure. Dialogue with the PEO or PM is important, particularly when quantities or timelines are changed in the POM because of funding changes. It is also important for sponsors to coordinate with their respective Deputy Assistant Secretaries of the Navy.

The SYSCOMs have dual allegiance in that they supply engineering and life-cycle management support to and have technical authority over acquisition programs through ASN (RDA). For all other activities, they report to the CNO. Thus, SYSCOMs may choose to raise their issues through ASN (RDA) rather than directly with OPNAV. Conversations should take place at all levels of management to avoid surprises on both sides. This means that resource sponsors, N80, and N8 leadership need to be in contact with relevant PEOs, PMs, and SYSCOMs.

Under Secretary of the Navy

The Under Secretary is charged with supporting the SECNAV, and, except in unusual circumstances, the Under Secretary's role in the PPBE processes of either Service varies by personality and SECNAV. In recent memory, the Under Secretary has played an outsized role in the PPBE process and in the Navy's Quadrennial Defense Review efforts. As Chief Management Officer, he or she could also have a large effect on manpower or structural levels in the POM. The Under Secretary is also the SECNAV's representative to the Defense Management Advisory Group, which adjudicates program and budget decisions at the OSD level.

When in a more active posture or provided a mandate by the SECNAV, the Under Secretary represents an additional battle rhythm of staffing and vetting that must be accommodated in the sometimes hectic schedules. As a result, plan appropriately.

Assistant Secretary of the Navy (Financial Management and Comptroller)

ASN (FM&C) oversees DoN's PPBE process for the SECNAV, including preparing the DoN POM as prescribed by SECNAV Instruction 5430.7R (a revision in staffing as this guide was drafted). In this context, ASN (FM&C) has a similar POM orchestration role for DoN and the SECNAV as the N8 has for the Navy and CNO. The main difference is that the CNO's influence and focus during POM development tends to be on planning, programming, and transitioning to budgeting, while ASN (FM&C)'s focus is the transition from programming and budgeting to execution.

As stated earlier, while ASN (FMB) is concerned with DoN program executability (among other metrics), ASN (FM&C) actually directs and manages the financial activities of the Department. It oversees and facilitates Navy and USMC interactions with OSD during OSD and congressional budget reviews, oversees DoN budget execution, and provides independent cost and budget analyses to the SECNAV, the acquisition community, the Services, and OSD.

Deputy Under Secretaries of the Navy

There are two Deputy Under Secretaries of the Navy prescribed by SECNAV Instruction 5430.7R (a revision in staffing as this guide was drafted). They are the Deputy Under Secretary for Policy and the Deputy Under Secretary for Management. In the PPBE context, the Deputy Under Secretary for Policy advises the Secretary and the Under Secretary of the Navy on key naval capabilities and concepts and is the author of the SECNAV guidance for the DoN POM. The Deputy Under Secretary for Management has domain for information technology business systems and management processes and would thus be interested in the Navy's PPBE process as it relates to that portfolio.

The Deputy Under Secretaries' other interactions with the PPBE process are usually interjections of oversight and requests for information at various times. Similar to an active Under Secretary of the Navy, these deputies impose additional work in both planning and response and must be accounted for to ensure that the milestones and deliverable requirements from POM serials are met.

Secretary of the Navy

The SECNAV—not the CNO or the Commandant of the Marine Corps—owns the DoN POM. The Secretary's authorities and opportunities to influence and adjust the Navy's program extend well beyond the Service Chiefs' in both scope (based on his or her Title 10 authority) and time frame (based on the SECNAV's typical interactions and the fact that he or she integrates and balances the Services' POMs for submission as a Department product to OSD). While some Secretaries have chosen to be largely hands off for the POM process, in recent memory, several SECNAVs have played much more-active roles.

Guidance from the SECNAV may not be explicit, so his or her speeches, testimony, and other communication venues may need to be examined to identify red lines and priorities. More recently, the SECNAV has not only been more active but has chosen to issue guidance that contradicts the Deputy Secretary of Defense. The rationale or relative merits of such a decision are beyond the scope of this guide. What is relevant to the Navy PPBE process is that (1) the SECNAV owns the POM and may choose a more or less active role, (2) related decisions are the Secretary's prerogative (that is, the SECNAV's authority is absolute), and (3) OPNAV AOs and leaders must account for the SECNAV's guidance in their planning and programming.

The bottom line: Expect the N8's guidance, as expressed in N80's POM serials, to reflect the CNO's coordinated interpretation of the SECNAV's guidance, unless otherwise directed. This is why an uncoordinated decision by a sponsor to depart from guidance in developing a POM submission carries risks.

External Stakeholders

Secretary of Defense and Deputy Secretary of Defense

Obviously, the Secretary of Defense determines what is in or out of the POM typically through the OSD staff, and he or she chairs the Defense Management Action Group. Generally, OSD CAPE or OUSD (Comptroller) determines the fiscal and programming guidance and sets the expectations for the POM and OSD's budget review. While resource sponsors have little to no influence over this part of the process, they should be involved in developing the justification for the Navy positions via the N8. Various under secretaries may ask for and receive POM briefs from Navy and Marine Corps staffs. The N8 produces those briefs, but N801 usually makes the presentations. The N801 branch is the Navy's voice and primary liaison with the OSD staff. However, it is anticipated that various staffs across OPNAV are in contact about issues as they arise. N80 needs to track issues carefully as the POM develops. After submission, it is N80's job to assemble all responses to requests for information, briefs, or clarification from OSD, combatant commanders, and the Joint Staff. Lastly, during program review, N80 is the lead voice in all matters with OSD issue teams to ensure alignment with CNO and SECNAV intent.

Note that the N8's role as external POM communications coordinator requires a different level of coordination now that the platform sponsors have been realigned under the N9, because those sponsors are now outside the N8 chain of command. Thus, resource sponsors need to proactively work closely with their N8 counterparts to ensure that appropriate subject-matter experts attend each engagement, depending on the audience, purpose, and venue.

Office of the Under Secretary of Defense (Comptroller)

OUSD (Comptroller) holds the final approval in the PPBE process. Issues are raised and decisions rendered through the Resource Management Decision mechanism. DoN can contest individual issues in writing, but it also has the opportunity to *reclama* (that is, request the reconsideration of a decision or change in policy) in person during the major budget issues session with the Secretary and Deputy Secretary of Defense. Generally speaking, for an issue to be adjudicated in DoN's favor at this stage should not be expected. The best posture for someone in OPNAV is to talk with his or her OSD counterpart to solicit support for championing the cause as desired (although this rarely occurs). Because ASN (FMB) normally orchestrates Navy's response in this process, information flowing from the Navy to OSD is normally expected to pass through FMB in this phase.

Office of the Under Secretary of Defense (Acquisition, Technology, and Logistics)

OUSD (AT&L) is OSD's chief acquisition executive office and has cognizance over DoD's acquisition programs. It also has cognizance over DoD's research, development, test, and evaluation appropriations. Its purview includes how the Service POMs affect the various acquisition programs that it oversees (usually ACAT 1D, but sometimes ACAT 1C as well). This office is also the proponent for science and technology, the industrial base, and environmental and energy programs. It is common for N80 to brief the Navy POM to OUSD (AT&L). Specific issues arising from changes in acquisition programs are addressed, and PEOs may be asked to attend. To coordinate interactions with OUSD (AT&L), sponsors should contact their respective Deputy Assistant Secretaries of the Navy.

Office of Cost Assessment and Program Evaluation

OSD's CAPE manages the review of military departments' and agencies' POMs during the program review cycle. At the time this guide was drafted, there were ongoing discussions about extending the time between the delivery of the OSD POM (back to the end of May) and the DoD budget (to the middle of September). The 2017 POM was a first step in this direction.

CAPE has historically been the major protagonist of the Navy in the POM cycle in that it raises capability, capacity, affordability, and executability issues to the Secretary and Deputy Secretary of Defense. CAPE challenges programmatic tenets and ideas and typically has the ear of the Deputy Secretary of Defense. It is important to work with N80 or N81, as applicable, to communicate ideas at that level, because both N80 and N81 have regular contact with their CAPE counterparts throughout the POM cycle and exchange information as needed. For example, N80 regularly discusses assumptions and interpretations of guidance with CAPE. Similarly, N81 coordinates major analytic assumptions, Analysis of Alternatives assumptions, and all planning scenarios and campaign analyses with CAPE counterparts, because any capability, capacity, or risk assessments based on analysis that is not consistent with approved assumptions and contexts will be rejected by CAPE.

Pricing and cost analysis are the purview of the cost segment of CAPE, and it will challenge pricing or cost estimates that differ from the acquisition cost position of DoN. Being aware of OSD studies and previous analyses is clearly important; if CAPE is doing a study in your sponsorship area, you need to be involved. CAPE usually produces military department fiscal guidance through contact with the Secretary of Defense and OUSD (Comptroller).

In all of these interactions, sponsor AOs and leaders should work through their N80 or N81 counterparts, as appropriate.

Office of the Secretary of Defense–Level Review Groups

Although names change every so often, the Deputy Secretary of Defense convenes a group of principal advisers to review POM and later budget issues to produce guidance for the military departments. Service under secretaries and vice chiefs represent the Services. Currently, this is managed through the Three-Star Programmers board and Deputy Secretary's Management Action Group for programming, budget, and review, with N80 or FMB serving as the OPNAV office of primary responsibility.

These are generally very important venues, and the reviews are often predictable. The bottom line for sponsors and N80 is to look forward based on history and recent signals and communications from OSD that highlight (or underscore) likely levels of interest. For some "troubled" or high-interest programs, there will be no question of whether they will be cut—only when. Work closely with your counterparts and be ready with appropriate information, narratives, and recommended attendees when the time approaches or events emerge. First and foremost, the information must be correct, provide coherent rationale and recommendations (as necessary), and be aligned with the Navy's approved narratives.

Chairman of the Joint Chiefs of Staff

The Chairman of the Joint Chiefs of Staff and sometimes the VCJCS can have views on the Service POM. To the extent that the Navy position can be explained to J8, it should be done by N80 and resource sponsor support. The Program and Budget Analysis Division of the VCJCS is the usual entry point.

The VCJCS is the Joint Chiefs of Staff representative on the POM and at budget meetings. He or she also represents the combatant commanders' viewpoints on POM and budget issues. Because N803 is the focal point for contact between OPNAV and the Joint Chiefs of Staff regarding JCIDS matters, N80 plays a somewhat larger role in deciphering acquisition-related issues likely to be raised by the vice chairman and J8.

The same point made earlier about coordination between sponsors and the N8 to ensure that appropriate subject-matter expertise is represented at engagements applies to Joint Chiefs of Staff interactions as well. When the issues and narrative center on the integration of capability, capacity, and risks, ensure that the appropriate representation from sponsors and N81 has been considered.

Congress

While Congress has no direct role in the POM development process, it is generally assumed that the staffers have access to various POM positions through weapon contractors and other external stakeholder organizations. The House and Senate Armed Services Committees authorize manpower levels and direct other issues, such as the retirement of platforms. DoN is aware of these issues, and a decision to push back is usually done at the SECNAV level in the POM. Of course, in the final analysis, the budget is approved by Congress as modified. While the POM is not delivered to Congress, portions of it usually leak to the principal staff members of important defense committees. Sponsors and N80 must be aware of these influences as they drive to balance the POM in the end game.

Key Relationships in the OPNAV PPBE Process

Chief of Naval Operations and the N8

The N8 represents the CNO in the POM and, with N80 and N81, provides the CNO's viewpoint in the POM deliberations. N80 and N81 usually agree, but programming and analysis can be two different things, and the fiscal versus warfighting capability-capacity-risk perspectives can drive organizations to different conclusions. Resource sponsors should not be surprised by that. Nevertheless, assume that the N8, N80, and N81 assumptions and positions reflect the CNO's guidance.

Depending on the CNO's business rules, he or she may be more or less inclined to use or require supporting analysis for framing and deciding major capability, capacity, programmatic, and risk issues. If less inclined toward analysis, this weakens N81's role, but it may also tend to weight advocacy or fiscal factors more heavily.

How the CNO wants to bring the POM together (i.e., integration and balance) depends on his or her business rules. The following are basically the three broad POM governance alternatives open to the CNO:

1. The CNO can defer to the N8, depending on the N8 to do most of the hard work in leading the DCNOs through most of the difficult choices (in alignment with guidance) and bring the POM to balance, at which point the CNO reviews the nearly finished product and makes adjustments at the end.
2. The CNO can interact with the N8, still depending on the N8 to orchestrate and integrate the POM, but with frequent CNO engagement at key points in planning and programming, and much more frequently at the end game.

3.   The CNO can reserve all or most integration authority to him- or herself, shifting more authority to the other DCNOs to advocate their positions, with the N8 mainly orchestrating the POM but not making major decisions.

The first two governance alternatives are most typical, historically. The three alternatives involve progressively more work for the N8. In fact, the third alternative practically requires the N8 to build two POMs—the one emerging from the collaborative process, usually billions out of balance as the end game approaches, and the other ready when OPNAV runs out of time and the SECNAV directs (or allows) FMB to lock the balance database for budgeting.

It is beyond the scope of this guide to judge the relative merits of these approaches. What is important is for the staff to understand which governance model is in play and how that will affect the amount and type of work needed to develop and deliver the POM. Again, the AO-to-AO and leader-to-leader relationships emphasized in this guide still matter regardless, and the staff's job is still to support the leadership, however it chooses to do business.

Secretary of the Navy and Chief of Naval Operations

The relationship between the SECNAV and CNO is critical to success in POMs. However, in the final analysis, the Secretary is dominant. Expect the CNO's positions and guidance to align with the Secretary's. Where they do not, expect those positions to have been coordinated at the executive level. The bottom line for the staff is to align with and follow the CNO's guidance, as known.

The N8 and Other Deputy Chiefs of Naval Operations

The N8 and the other DCNOs are equals, but the N8 also represents the CNO's need to deliver a coherent, balanced POM that reflects the SECNAV's and OSD's guidance. Because the N8 does not own a program (N84 and N89 notwithstanding), he or she tends to focus on the integrated product. But because of this focus, the emphasis also tends to be fiscal and risk balance, and the time frame of reference is typically the FYDP.

The 2011–2012 OPNAV realignment that established the N9 shifted program-specific manpower and readiness programming authorities back into the platform sponsors. But that realignment also was intended to free up N8 executive bandwidth by providing another three-star (N9) who could devote more time to longer-range force development planning while the N8 focused on integrating, balancing, and communicating the Navy Program. This means that the N9 and N2/N6 now have a larger, more independent share of the burden of planning the future Fleet. This, in turn, fosters a more distinct tension between views of the Navy Program—from a fiscal-programmatic perspective on one hand and from a perspective focused more on capability, capacity, and outcome on the other. The balance between these two perspectives will likely continue to shape the roles, relationships, and interactions between the DCNOs in coming POM cycles.

The dominant factor that shapes the relative roles between the N8 and the other DCNOs is, of course, the CNO. If the CNO prefers more-exclusive, close-knit decisionmaking, this tends to mean a first-among-equals role for the N8, which may affect how open the N8 is and when information flows to the DCNOs. If the CNO is more inclusive, DCNO roles tend to grow and influence opportunities tend to be more frequent.

N80 and N82/FMB

N80 and N82/FMB usually have a close relationship and cooperate with each other, although they sometimes interpret guidance differently and may take different actions on similar issues—if for no other reason than FMB touches the POM after N80 and conditions (fiscal, policy, political, and so on) sometimes change as programming transitions to budgeting. Another cause of potential differences is that N80 is attuned to the CNO, through the N8, while FMB is more attuned to the SECNAV.

A resource sponsor should not assume that if one loses an issue in the POM, he or she can win it back in the budget. However, it has happened. Nonetheless, the two processes are quite different and respond to different stimuli. The POM responds to guidance from the CNO and SECNAV, and the budget tends to respond to guidance from the SECNAV, OSD, and Office of Management and Budget. The POM responds to resource sponsors, and the budget responds to BSOs, or claimants. The POM is concerned with the FYDP, and the budget usually is the first year of the FYDP.

FMB budget analysts also hold meetings with the SYSCOMs, PEOs, and Fleets, focusing on pricing, programmatic risk, and executability of high-interest programs and accounts. Resource sponsor interactions with those stakeholders are usually more intense during planning, early phases of programming, and execution, and they usually focus on high-priority requirements and gaps.

To avoid surprises during budgeting that were not accounted for during programming, maintain ongoing relationships with the organizations that have to execute the appropriated resources. For platform sponsors, that means TYCOMs, PEOs, and SYSCOMs. For resource sponsors, it means the people and establishments affected by the programs you resource. For N80, it means sponsors and FMB.

N80 and Resources and Requirements Sponsors

N80 and resources and requirements sponsors have a push and pull relationship, by design. N80 is responsible for orchestrating and integrating the POM and keeping it on schedule. A common misperception is that N80 staff are trying to hide what they are doing (which is rarely the case, and when it is, those actions are usually directed). Usually N801, N80, and the N8 are running quick-turn efforts to respond to CNO or

SECNAV (through FMB) tasking. They do not normally have the time to seek you out and tell you what is happening. Proactive engagement with them usually gets you the current, accurate (albeit, potentially volatile) information you seek.

By tradition, N80 is supposed to be mostly open about the guidance it is working under, the assumptions it is using, and the decisions it is making. This includes the strong, built-in incentive to reduce the amount of work it takes for N80 to integrate and deliver the POM.

There is a pragmatic tradition for AO programmers working for sponsors and N80 to keep in close touch, particularly as the time for the end game (or final balancing of accounts) gets close. When this horizontal AO-to-AO coordination is good, the POM tends to move through the process more easily, and with fewer surprises. However, a recurring lesson learned is that this lower-echelon coordination is much less effective (and more limited in scope) when vertical and upper-echelon coordination breaks down (or is less diligently maintained). There have been POMs in which major surprises have occurred at executive briefings, only to find that the working-level staff was more or less completely aware of the issues.

This is another example of staff failure, and one way to reduce its likelihood is to focus on communication (whether in agreement or not) and maintain working relationships up the chain and across stakeholder organizations in the interest of the common objective—informed planning and decisionmaking aimed at delivering a coherent, balanced POM in alignment with the CNO's guidance.

The bottom line for AOs and ROs is that if you are not a programmer, maintain a working relationship with both your own division's POM shop and your bullpen counterpart (i.e., the N80 programmer responsible for the accounts that resource your programs). The bottom line for leaders is to actively maintain a free flow of information and awareness both within your own organization (bidirectional between working and executive levels) and among your executive stakeholder counterparts.

CHAPTER FIVE

Best Practices for Succeeding in the Navy PPBE Process

This chapter offers practical advice to all stakeholders on how to succeed in the PPBE process and suggests how to define and measure success. It discusses the broad principles for organizing and planning to improve the chances for success. Given the nature of the processes involved in the POM and the budget, it also addresses how to set and manage expectations and how to avoid problems and surprises. It outlines typical sources of uncertainty and instability and how to deal with them. Inevitably, problems will crop up over the course of the POM cycle. This chapter suggests what to do if that happens, offers some lessons learned, and provides examples of what has worked in the past—and, equally important, what has not.

The authors have tried to frame the following guidelines in universally applicable (change-agnostic) terms—that is, applicable even if fundamental PPBE processes and key stakeholder roles outlined in previous chapters may have changed.

Understand Whose POM It Is

As mentioned earlier, the most important thing for an OPNAV stakeholder to remember is that (unless otherwise directed by the CNO), for OPNAV staff, the POM belongs to the CNO. And for the CNO, the POM belongs to the SECNAV. Title 10 authorities and relationships between the SECNAV and CNO notwithstanding, OPNAV's job is to deliver a timely, integrated POM whose priorities, balance, and rationale are coherent with the CNO's guidance and intentions through the N8 to FMB.

Understand What Success Means and How to Measure It

Although there is no doubt that a successful POM can be defined and measured at multiple levels, the most important measure of success is that OPNAV did its job as just defined. Nevertheless, there are other secondary measures that should be considered, as long as it is acknowledged that all other measures derive their meaning from and are trumped by the CNO's guidance.

Some readers may be surprised that success is not defined as delivering a POM that adequately resources the critical capabilities that the Navy needs, but all POMs try to do that to some measure. However, leaders' prerogatives are to shape priorities, make (or defer) decisions, and (explicitly or implicitly) accept more or less risk. Therefore, assuming that the overarching objective is to deliver a POM that enables a "forward, engaged, and ready" Navy, consider some more-tactical characteristics that would make a POM submission successful, such as the following:

- is on time
- is integrated—vertically and horizontally
- unambiguously expresses the CNO's priorities
- aligns with the CNO's guidance on balancing requirements, resources, and risk
- is coherent with the CNO's POM story line (or theme).

From these characteristics, it is straightforward to derive subordinate measures of success.

With respect to specific requirements of interest, a useful measure of success is to track proposed outcomes with actual decisions as they flow through the POM cycle, where success means maximizing favorable decisions that enable desired outcomes. Note that this approach must be caveated with the observation that few decisions are final or irreversible. Given the number of downstream decisionmakers whose rule sets may differ from that of any given stakeholder, success in this context means successfully negotiating each of these key decision points—usually over several years, from planning through execution.

The bottom line here is that measuring success is not as simple as how well a stakeholder protected the resources of his or her programs. In fact, if everyone on the staff followed that metric, the POM would be an incoherent, out-of-balance failure. The tension between advocates and integrators in OPNAV's PPBE process is by design, but the process works best when all stakeholders understand where they and their programs fit into the process and portfolio and how they work together to deliver the Navy the nation needs.

Understand the Navy Program and Where Your Portfolio Fits

In this context, the Navy Program is, broadly, the whole portfolio outlined in the annual Navy Program Guide, posture statement, and classified POM serial, as described earlier. The better that stakeholders understand what goes into the Navy Program and where and how their particular program or portfolio of programs fits into it, the easier it will be to understand and follow the guidelines leading to a successful POM. Where and how a portfolio fits depends on both historical (often traditional) and circumstantial contexts.

In addition to the typical published messaging products—for example, the current maritime strategy document and the Navy Program Guide—there typically is a POM or portfolio story line (a "narrative" or "theme"; see Chapter Six) that succinctly articulates how the Navy's requirement priorities and corresponding investments contribute and relate to Navy, DoD, and national strategy. As of 2015, the N8 is the lead for the POM story line, but developing it was more straightforward (and convenient) when the platform sponsors were aligned under the N8. With the establishment of the N9, the process and character of the annual narrative likely will evolve.

The bottom line here is the same recurring theme throughout this guide: Stakeholders should participate in developing the story line (as appropriate) and always know and be ready to articulate where and how their programs fit into the narrative.

Understand the Requirements Space

Recalling the definitions outlined in Chapter Two, it is critical that stakeholders understand what requirements are in play during each phase of POM development and who owns each requirement (i.e., which stakeholders are responsible for resourcing them); in addition, stakeholders must understand that those requirements competing for resources have been both documented and validated. These are prerequisites to understanding how the issue might be addressed in the POM, which stakeholders and decisionmakers likely will be involved, and what influence any AO or other stakeholder is likely to have on the outcome. The issue of ownership is often ambiguous, and because capability, capacity, performance, and programmatic requirements all translate to bills that someone may have to pay, stakeholders often seek to keep ownership ambiguous (unless they can help pin responsibility on some other organization). It is also common for many requirements to have advocates who are not responsible for resourcing them. Examples include the N1 for (most) manpower requirements, the N4 for (most) readiness requirements, and platform or system sponsors for various C4ISR systems critical to their effects chains.

Stakeholders also need to maintain awareness of the requirements life cycles of their programs. This means being aware of development and reviews of formal requirements documentation, gate reviews, Defense Acquisition Boards, and configuration control boards.

The key point here is that stakeholders must understand what requirements are in play; their validation, configuration, and resourcing status (both historically and future); and who is (or may be) responsible for advocating and resourcing them.

Understand the Trade Space

There are both traditional and circumstantial trade-space priorities and relationships among future capability, current readiness, force structure, and personnel requirements that tend to guide leaders' intentions and decisionmaking.[1] It is critical for stakeholders to understand the priorities in play during a given POM cycle, and it is equally critical to understand where their portfolio sits in that trade space. This is not just to manage expectations of outcomes; it is also to help define what work is useful and what work is, in the end, just a waste of time. Stakeholders should be mindful of where their portfolio fits in defining objectives, developing a plan, prioritizing the work that gets approved, and deciding when to do that work.

Another issue relates to overall resource priorities for which there are clear differences in advocacy and support. For those portions of a stakeholder's portfolio that do not have strong external advocacy (say, readiness-related programs affecting maintenance and training), it may be necessary to give these programs higher internal priority in the portfolio during the POM build and take some risk in what would be externally characterized as higher-priority programs to the Navy, because those likely will receive higher attention and garner additional support (within the division or directorate or at the CNO or SECNAV level).

The bottom line here is that OSD, SECNAV, and CNO guidance may specify (or imply) clear (or ambiguous) priorities between force structure (procurement and manpower), future capability (modernization), readiness (e.g., training, maintenance, logistics) and personnel (e.g., manning, medical). It is the staff's job to understand how leaders have chosen to interpret higher-level guidance, what the current guidance is, and how that guidance and the specific configuration of priorities between force structure, future capability, readiness, and personnel will affect POM development and trade-space priorities for the sponsors, N8, and CNO.

Leverage the POM Schedule and Deliverables as Planning Tools

Beyond the obvious advice to know the PPBE processes and follow the POM serials (see Chapter Three), less obvious is how this understanding should affect day-to-day planning. Even in periods of uncertainty, a lot of tasking and questions are predictable and can be anticipated in the context of the phases, events, and deliverables described in the serials. It is always a good idea for a stakeholder to exercise his or her organization's relationship with N80.

The point is that much tasking, preparatory work, and even many "surprises" can be anticipated simply by inferring them from what likely will be going on at that time

---

[1] *Trade space* can be defined as the program and other parameters and characteristics required to satisfy performance standards. Leaders must make trade-offs in cost, risk, performance, and so on when defining the trade space.

in the POM schedule. This is true even during periods of high uncertainty because many calendar-driven and process-driven POM activities continue to grind along, despite churn between OSD, Congress, and the White House.

The key knowledge that typically unlocks this ability to anticipate, however, is how well stakeholders understand the state of their program or portfolio and how it has played and likely will play in each phase of the POM process. The stability and maturity of requirements, how well a program is performing, the resource outlook, alignment with leadership, and so on are all factors in the process.

Understand Stakeholder Rule Sets and How They Change over Time

At numerous points in the PPBE process, normally at the transition between phases (e.g., strategic planning to guidance, requirements assessment to POM review, planning to programming, programming to budgeting), the "lead" passes from one key POM stakeholder to another. Each stakeholder is guided by a different rule set affecting his or her point of view, motivations, assumptions, value propositions and metrics, and so on. These rule sets affect what matters and the likely outcomes at key decision points in the process during the time that person has the lead (which is typically his or her window of maximum influence on outcomes). The bottom line here is that these rule sets affect how relevant an AO's or other stakeholder's requirements are to the decisions facing the lead stakeholder, as well as how that person is likely to evaluate the requirements in the context of competing requirements and the Navy Program as a whole. Examples of the leads at various phases in the PPBE process include the following:

- the N3/N5 during development of strategic guidance
- N81 during development of the analytic agenda and reporting of studies and assessment results
- N80 during development of the POM schedule, development of fiscal and programming guidance, and integration
- sponsors during the SPP builds and out-briefs
- the N8 during integration of the SPPs and the end game
- FMB during and following the transition to budgeting.

Because these phase transitions represent shifts in supporting and supported relationships between stakeholders, the lead shift can change how a program or requirement is perceived relative to others. This can put a stakeholder's portion of the POM at varying degrees of risk. For example, for any given time,

- not every stakeholder may have participated in the decisions leading up to the current state of assumptions, priorities, decisions, or rationale

- details may get lost and interdependencies, if not explicit, may be forgotten or not recognized
- the way an issue or program is presented or articulated may change between phases
- how an issue is evaluated almost certainly changes between phases.

The key point here is that, for stakeholders to have some confidence in their chances for success in each new phase, they need to create and communicate solid analysis and logic for their positions.

Understand and Exercise Your Relationships with Other Key Stakeholders

A consideration that frequently leads to misunderstandings or disappointment is failure to understand and exercise ongoing working relationships between stakeholders. This recurring theme has been emphasized throughout this guide but will be restated here more emphatically: Stakeholders cannot hope to be successful in the PPBE process without routine, ongoing communication and interaction with other stakeholders

- within their chain of command (vertically and horizontally)
- across OPNAV (horizontally, with counterparts)
- with key external Navy and non-Navy stakeholders
- at both working and executive levels, such as CAPE or OUSD (AT&L).

This interaction is more than simply maintaining awareness and communicating, and it absolutely must be independent of the level of agreement between participants. In most cases, it involves deliberate participation in one or more stakeholder processes to maintain the ability (when necessary) to exert timely influence on an issue or decision. This influence can be aimed at either working toward positive outcomes or avoiding negative ones, but it is usually contingent on both being aware of the issue and having standing to influence it. Both circumstances require ongoing commitment to participation.

The quality of the POM is so dependent on the effectiveness of these relationships that the ability to maintain them should have significant influence on leaders' prioritization of how they spend their day, what work is important, and what work should be avoided, deferred, or divested. Fostering and protecting key stakeholder relationships should be a high priority in all work-plan development and workforce management decisions.

Manage Expectations

The POM process is engineered—from top to bottom—to encourage tension between stakeholders and facilitate incremental change in current and future military capability and capacity. Stakeholders with aspirations to facilitate revolutionary changes in the Navy Program should prepare themselves for disappointment. That said, it is still important to set realistic goals between inconsequential and revolutionary change and then set priorities and scale POM activities appropriately.

Success should be aligned with leadership objectives. If stakeholders focus on protecting their own equities, it may work occasionally, but it is unlikely to align with the Navy's objectives or guidance, and, collectively, it is a self-defeating expectation. Early in the process, there is a time and place to influence corporate positions. By the time SPPs are submitted, it is usually a bit late.

Pay Attention to CNO Guidance, Testimony, Speeches, and Decisions

Treat everything that the CNO publishes or says in testimony, speeches, and decisions as guidance to be followed (unless otherwise directed). Failure to maintain a good understanding of the message embodied in what the CNO says almost always leads to problems. If these problems become surprises (i.e., leadership thought things were going one way and a stakeholder did something else instead), the outcome will not be good. It is the staff's responsibility to align with the CNO's guidance and priorities.

More broadly, whether intentional or not, the collection of messages emanating from the CNO's office over time forms a story and generates themes that should affect POM priorities, how the POM is built, how it is balanced, and how it is defended. Ideally, each stakeholder should not need to interpret all of this independently. That said, it is important for each stakeholder to recognize, account for, and act on the portions of the CNO's message that speak to their requirements or programs.

When OPNAV develops a coherent POM or portfolio narrative, the process of understanding and aligning becomes much more straightforward.

Follow the Guidance

It is important for all stakeholders to follow appropriate guidance, but, at times, the guidance issued by OSD, the SECNAV, and the CNO may be ambiguous or inconsistent. Likewise, a portion of the chain of command may choose to take a different path. The causes, motives, and risks associated with these circumstances are beyond the scope of this guide. Whatever the circumstances and causes, a stakeholder should follow the guidance he or she has been given and have a plan if the CNO has made it clear what he or she wants but the stakeholder's chain of command decided to deviate from that path.

That said, the one DCNO normally guaranteed to align with and enforce the CNO's guidance is the N8. So, regardless of the various uncertainties that may plague a given POM cycle, rest assured that if an SPP turned in by a sponsor does not align with the CNO's guidance, N80 will adjust it. In other words, stakeholders who fail to follow the CNO's guidance should expect changes to their POM submissions.

There may be reasons not to follow guidance (or to choose one source of guidance over another). Under some circumstances, it could be an effective strategy, but if stakeholders are going to do it, they should always coordinate with Navy leadership. In addition, deviating from guidance virtually compels stakeholders to prepare two versions of the POM database—one aligned with the advocated guidance and one ready to apply when (or if) the N8 or FMB rejects the submission and directs compliance. Sponsors who fail to have this hedge position prepared effectively hand their voice to the N80 integrators, who, on a short timeline, must take apart and reassemble the affected programs to comply with guidance. N80 or FMB will balance the programs as directed by leadership, with or without sponsor input.

Understand How and When to Solicit and Interpret Decisions

In the weeks following a CNO Executive Board, a frequent refrain is, "The CNO loved my brief; why did he cut my program?" This is a manifestation of a broader tendency among senior leaders who must make trade-offs between or across programs; these leaders often feel they must hold cards close to the chest and cannot signal decisions before their time. If a decision forum is not specifically designed to solicit a decision for the record, do not expect to learn the outcome in the meeting (or even shortly afterward), and do not try to infer it from how well a presentation seemed to be received.

Also, it is a waste of stakeholder and executive bandwidth to solicit a decision out of sequence or one that is highly interdependent with undecided or uncertain requirements, programs, or issues. For example, asking leadership to make an early decision about a platform that will have major implications for operating costs or infrastructure is likely futile. The most recent (as of this writing) OPNAV realignment that restored program-specific manpower and readiness resources to the platform sponsors was intended specifically to address and prevent some of these issues, but dependencies with other parts of the Enterprise and other portfolios also must be considered.

Another consideration is to understand who the decisionmakers are and how they make decisions. This includes how they like information presented to them, the types of content they expect to see, and the other stakeholders whose advice they typically solicit.

Often, the reason for disconnects between presentations and outcomes is failure to recognize and follow guidance. Leadership may be unaware of the consequences of what has been presented until they actually see the numbers.

Avoid Surprises

Surprises during the POM cycle are usually not a good idea unless engineered for a specific reason (usually above the pay grades of the audience reading this guide). Establishing and maintaining transparency throughout the process is usually a good way to avoid surprises, but surprises can also occur when two stakeholders interpret decisions or guidance differently. Good guidelines for avoiding surprises include the following:

- Maintain awareness of critical issues and activities throughout the cycle.
- Be more inclusive in order to keep more eyes and ears open (for example, involve stakeholders in planning and decision meetings).
- Verify what is going on with reviews at key stages in the POM cycle.
- Communicate often and at multiple levels (executive and working).

In line with these guidelines, it is always a good idea to stay linked with the N8 and N80 at the working and executive levels throughout the process. Doing so almost always requires coordination within a chain of command to ensure that everyone knows who is responsible for a particular program or phase (usually the N8) and within what battle rhythm.

CHAPTER SIX

Planning for Success

With the best practices from Chapter Five in mind, we now outline ways to put them into practice.

Have a Plan

As should be clear from this guide by now, waiting for direction (either from a direct chain of command or from the N8's serials) is guaranteed to put a stakeholder behind schedule and put his or her portfolios at risk. Be proactive.

As has been said, even in conditions of high uncertainty, many universal aspects to building the POM still take place, no matter what the process specifics or what roles certain leaders have. Suggestions for forming a plan include the following:

- Understand how the upcoming cycle might change and what can be done about it.
- Participate in training (if new to the process, take training; if not new, train others to build some depth).
- Do some research into the
  - relevant history of applicable policies and decisions affecting assumptions and the trade space
  - current guidance (if not already known).
- Understand a portfolio by reviewing its programs (leaders should ask staff to present the programs, which is a form of training for both the leader and staff), including the key drivers and trends, such as requirements, schedule, costs, and risk.
- Understand the perception around the portfolio (for example, from discussions with other stakeholders, past decisions, opinion pieces, and position papers).
- Decide how to develop and exercise portfolio relationships.
- Develop (or align with) a POM strategy (normally in collaboration with peers as part of a division, directorate, or staff-wide process).

Protect the Ability to Plan

Because of how OPNAV works, the most-talented people typically are already over-loaded working the hot topics or processes. This tends to create a bias toward reactive activities at the expense of more-deliberate planning and execution. The types of planning that are needed at critical phases in the PPBE process require varsity talent, so stakeholder leadership needs to protect the bandwidth of personnel and shelter them from the daily reactive tasking of the Pentagon to allow them to plan. This pays dividends in the following areas:

- anticipating and prioritizing questions, taskings, and responses
- doing horizontal integration
- building a POM story line that shows how the program or portfolio aligns with guidance and fits together.

The main point here is that leaders must make time and people available for planning, or else it will not happen. The less planning that happens outside the N8 office, the more stakeholders yield their voices to the N8's staff and the more influence the N8 exercises over the final configuration and balance of the POM.

Develop a Battle Rhythm

Maintain a recurring drum beat of engagement and situational awareness (at the executive and working levels) in the context of the guidelines outlined earlier. In the battle rhythm, account for N80, N81, directorate integration staffs (e.g., the N2/N6F and N9I), and appropriate TYCOMs, PEOs, and SYSCOMs.

A battle rhythm helps focus on what is important when the inevitable distractions arise, often at the worst possible times. If done correctly, it keeps stakeholders informed and aligned and facilitates the types of communication and collaboration needed to stay aligned with guidance and avoid surprises.

Manage Participation and Engagement

Stakeholders should be deliberate and pragmatic about how they treat the activities around them. The following framework may be useful in helping decide where and when to devote time and effort, given all of the things that must be done to generate and defend the POM.

Stakeholders should take stock of the various activities around their organizations and apply the following tests of relevance and consequence to each:

- *Relevance level 0.* This activity is (and will continue to be) of no consequence; no matter what it does (good or bad), it will have no (or inconsequential) effect on what we care about. Therefore, ignore it.
- *Relevance level 1.* This activity could be important (either as a help or a threat), and we should pay attention to what is going on. Therefore, observe it.
- *Relevance level 2.* This activity is likely to be important (either as a help or a threat), and we need to be seen as a stakeholder and involved enough to ensure that we have time to react (one way or the other). Therefore, attend the meetings, briefings, and so on.
- *Relevance level 3.* This activity is likely to be important enough (either as a help or a threat) that we need to shape the outcomes (one way or the other). Therefore, participate in the process and devote people and resources to supporting (or defeating) the outcome.
- *Relevance level 4.* This activity is so critical to our success (either as a help or a threat) that we need to take steps to ensure the outcomes (one way or the other). Therefore, control it; do what is necessary to adopt, take the lead in, or kill the effort.

The point of this framework is, obviously, to be deliberate about how to engage (because time and human resources are precious) and ignore, maintain awareness, or engage as appropriate to the value of the relationship to a stakeholder's portfolio or activity.

Develop Coherent Justification

It is always a good idea to develop clear linkages among the requirements, resources, capabilities, capacities, costs, and risks. Doing so can pay dividends in a stakeholder's relationships with other requirements and resources stakeholders on whom his or her programs may depend; in the stakeholder's interactions with N80, whose staff will better understand how the programs fit together and with the rest of the Navy Program; in the stakeholder's interactions with N81, which will provide an operationally coherent, Fleet-level linkage among capability, capacity, and risk; and in the development of the stakeholder's SPP, because it will facilitate telling the story that leaders need to hear.

Proactively Coordinate Executive Schedules

Know the dates of all events in the PPBE cycle that afford access to Navy and DoN leadership. Influencing the schedules of these executives and encouraging or ensuring their availability for decisionmaking are the most important management actions affecting the efficiency of the OPNAV staff during the POM cycle (particularly during busy phases of the staff-year). Prior to the start of each phase (assessment, POM build, POM defense, Navy budget review, or OSD budget review), N80 must schedule a sufficient number of opportunities with the leadership to effect closure on the issues (i.e., to balance the issues).

As part of this process, avoid starting meetings in the late afternoon, especially approaching the end game. People who have worked late into the night make mistakes. This can hurt product credibility and make even more work for the staff cleaning up afterward.

Develop the POM Theme

Developing and articulating a thematic structure (that is, the story line or narrative) for decisionmaking is a powerful tool for effective POM-building. Develop this thematic structure with stakeholders and leadership early in the cycle. Later, the theme will serve as a guide for structuring the defense of the POM. Specific program guidance, available from leadership decisions and planning guidance, details what must be funded. Often missing is a structure for deciding on the margin what should be added or cut to balance. Fair share methods, such as sponsor share, have their place but do not ensure a coherent, capability-based end product. Giving N80 a set of standing guidance that consists of straightforward decision rules simplifies the analysts' tasks and promotes compliance with the POM theme. Therefore, the POM theme and its associated decision rules should serve as both an input and an output of the POM-building process.

Ensure that the POM theme or narrative can effectively communicate to all intended stakeholder audiences. Do not assume that the only important stakeholders are defense programmers. An effective POM theme has appropriate elements of warfighting, capabilities, capacities, risks, priorities, and programs, as well as a coherent way to weave these together.

Be Prepared to Use Estimated Controls Until Well into POM Development

Although it is understandable that sponsors need firm resource controls on which to begin POM development, it is very rare for OSD (and therefore, the N8) to release them before POM development timelines begin. Sometimes, the guidance is simply not yet available because of issues between the administration and Congress. It is also understandable that sponsors would wonder how they are supposed to prioritize and balance their portfolios when they do not know how much money will be allocated to each of their funding lines. Therefore, sponsors need to use estimates, which underscores that sponsor relationships with N80 are important.

Because significant changes in the major accounts are rare from year to year, rough estimates can be based, to some extent, on prior-year controls and known trends. DCNO integration divisions (e.g., the N2/N6F and N9I) should coordinate with N80 on developing sensible control estimates based on leadership guidance, and these have to suffice. The key point is for both sponsors and N80 to know what estimates are being used so that programmers can take this reference as a common baseline in their planning while positions and guidance become more certain. This communication should minimize surprises and clarify the adjustments that need to be made when controls are finally assigned.

Coordinate Controls with Anticipated Assessment Results

The controls specified in serials can be more effective if shaped in coordination with the major POM assessments (e.g., N81's FEA, the N4's Integrated Readiness Assessment). It may be necessary to rely on early or preliminary study results because final results do not get reported until the middle of the POM build. It is usually unlikely for the major conclusions from these studies to change, and the benefits of considering these results are that controls can be shaped to help facilitate the types and magnitudes of trade-offs that sponsors need to make to reach balance. This is particularly important for manpower and readiness accounts now that sponsors have resource authority over those accounts for their major platforms. By anticipating the types of trade-offs that may need to be considered, controls can be more or less specific to accommodate or shape those trade-offs, depending on CNO, SECNAV, or OSD guidance.

Coordinate How the POM Will Be Defended

Try to anticipate which elements of the POM may be left to individual sponsors to defend—a decentralized approach that can reduce staff workload and increase the speed and number of issues being addressed—and which should be defended centrally, within the N8's purview. This is less about loyalty and more about consistency and staying on message. When it is critical that defense of the POM (or some elements thereof) be coherent with specific Navy positions, consider who should lead that effort (typically a representative within the N8's office) and ensure that those arrangements are communicated to all stakeholders.

CHAPTER SEVEN
Conclusion

The PPBE process is complex, with lots of relationships, traditions, and unwritten rules. It takes the average newcomer a year to be conversant and marginally capable, plus a second year to begin exercising and demonstrating journeyman skills that permit leaders to trust the results. As of 2015, it was the authors' observation that nearly three-fourths of the military staff in OPNAV had fewer than two POM cycles of experience.

This guidebook was aimed at AOs, branch heads, newly assigned flag officers and executives in OPNAV, and outside stakeholder organizations with an interest in the OPNAV PPBE process and how the Navy executes it. It has tried to represent the perspective of requirements and program advocates, as well as requirements and resource integrators. It has tried to highlight those aspects of the PPBE process, products, stakeholders, relationships, interactions, and best practices needed for readers and their organizations to be successful individually and as a staff.

It is the authors' hope that this guide can facilitate and improve the staff's ability to work together toward the common purpose of delivering a coherent, balanced POM in alignment with leadership guidance.

Bibliography

Assistant Secretary of the Navy, Financial Management and Budget, *DON Budget Guidance Manual*, Washington, D.C.: Department of the Navy, July 2013.

———, Congressional Information Management System, web portal, Washington, D.C.: Department of the Navy, 2015a.

———, Program Budget Information System, web portal, Washington, D.C.: Department of the Navy, 2015b.

Assistant Secretary of the Navy, Financial Management and Comptroller, "Financial Management and Comptroller," web page, Washington, D.C.: Department of the Navy, 2015. As of June 13, 2016: http://www.secnav.navy.mil/fmc/Pages/home.aspx

Defense Acquisition Portal, "Acquisition Process—Big 'A' (Acquisition) Process Concept and Map," web page, Defense Acquisition University, undated-a. As of June 13, 2016: https://dap.dau.mil/aphome/Pages/Default.aspx

———, "JCIDS Process," web page, Defense Acquisition University, undated-b. As of March 15, 2015: https://dap.dau.mil/aphome/jcids/Pages/Default.aspx

Department of the Navy, "Introduction to PPBE, Action Officers Course," courseware slides, Washington, D.C., 2015a.

———, "Introduction to PPBE, Executive Level Course," courseware slides, Washington, D.C., 2015b.

Department of the Navy Program Information Center, *Programming Manual*, Washington, D.C.: Department of the Navy, March 1985.

DoN—*See* Department of the Navy.

Office of the Chief of Naval Operations, *Office of the Chief of Naval Operations Organization and Operations Manual*, OPNAVINST 5430.48E, Washington, D.C.: Department of the Navy, November 2011.

Office of the Chief of Naval Operations, Programming Division, POM-17 Serials 1, 2, 3, 4, 4A, and 4B, Washington, D.C.: Department of the Navy, September–December 2014.

OPNAV—*See* Office of the Chief of Naval Operations.

OPNAV N80—*See* Office of the Chief of Naval Operations, Programming Division.

U.S. Marine Corps, Department of the Navy, and U.S. Coast Guard, *A Cooperative Strategy for 21st Century Seapower*, Washington, D.C., March 2015. As of June 13, 2016:
https://www.uscg.mil/seniorleadership/DOCS/CS21R_Final.pdf

USMC, DoN, and U.S. Coast Guard—*See* U.S. Marine Corps, Department of the Navy, and U.S. Coast Guard.